Spatial Intelligence for a Greener Planet

AI Applications in Environmental Pollution Mapping, Analysis and Mitigation

Editors

Kuldeep Chaurasia

School of Computer Science Engineering and Technology
Bennett University, Uttar Pradesh

Pradeep K. Garg

Civil Engineering Department
Indian Institute of Technology Roorkee
Roorkee, Uttarakhand

Arun P. V.

Indian Institute of Information Technology
Sri City, Tirupati, Andhra Pradesh

Yingwei Yan

Department of Geography, National University of Singapore
Singapore

CRC Press
Taylor & Francis Group
Boca Raton London New York

CRC Press is an imprint of the
Taylor & Francis Group, an **informa** business

A SCIENCE PUBLISHERS BOOK

First edition published 2025
by CRC Press
2385 NW Executive Center Drive, Suite 320, Boca Raton FL 33431

and by CRC Press
4 Park Square, Milton Park, Abingdon, Oxon, OX14 4RN

© 2025 Kuldeep Chaurasia, Pradeep K. Garg, Arun P. V. and Yingwei Yan

CRC Press is an imprint of Taylor & Francis Group, LLC

Library of Congress Cataloging-in-Publication Data (applied for)

ISBN: 978-1-032-69998-1 (hbk)
ISBN: 978-1-032-71828-6 (pbk)
ISBN: 978-1-032-71832-3 (ebk)

DOI: 10.1201/9781032718323

Typeset in Times New Roman
by Prime Publishing Services

Preface

In recent decades, the intersection of Geospatial Technology (Remote Sensing + GIS + GPS), Artificial Intelligence, and Environmental Science has transformed our understanding and management of pollution. This book explores the multifaceted applications of these technologies in mapping, analyzing, and mitigating environmental pollution across diverse landscapes and contexts.

Each chapter investigates specific methodologies and case studies that emphasize the transformative power of geospatial approaches and AI-driven solutions. Exploring the use of AI techniques such as machine learning, data analytics, and computer vision, the book illustrates their application in monitoring, predicting, and mitigating pollution sources. The book delves into various types of pollution including air pollution, water pollution, soil contamination, and thermal pollution—highlighting their detrimental effects on human health, ecosystems, and the planet. From groundwater quality assessments to urban sprawl modeling, and from spectral unmixing for water body pollution assessment to decision support systems for comprehensive pollution management, this book offers a comprehensive view of the evolving field. By drawing upon real-world case studies and cutting-edge research, we demonstrate how AI and geospatial technology can revolutionize pollution mitigation strategies.

The book chapters are authored by experts and researchers at the forefront of their respective domains, providing insights into groundbreaking strategies, challenges encountered, and future directions. Together, they illuminate how these cutting-edge technologies are reshaping environmental intelligence and enhancing our ability to safeguard the planet for future generations.

By the end of this book, readers will appreciate the synergy between AI and geospatial technology and their pivotal role in creating sustainable solutions for pollution control. This book is intended for researchers, policymakers, practitioners, and anyone interested in the key role of Geo-AI in addressing the complex challenges posed by environmental pollution. It serves as both a reference and a call to action, advising collaboration and continued innovation in our collective pursuit of environmental sustainability.

Acknowledgement

I am deeply grateful to all those who have supported and contributed to the completion of this book, "Spatial Intelligence for a Greener Planet: AI Applications in Environmental Pollution Mapping, Analysis, and Mitigation."

First and foremost, I extend my heartfelt gratitude to my parents for their unwavering support and encouragement. I also wish to thank my wife, Arti Chaurasiya, and my daughters, Advika and Aavya, for their love and continuous support throughout this book project. Their steadfast belief in me has been a constant source of motivation. I extend my heartfelt gratitude to my mentors, colleagues, and friends for their continuous inspiration and support. Their insights and advice have been invaluable in shaping the direction and content of this book.

I am especially thankful to my fellow editors for their invaluable contribution and collaboration. Their dedication and expertise have been crucial in bringing this book to fruition. A special thank you goes to all the chapter authors for their contributions. Their hard work and commitment have enriched this book with diverse perspectives and profound knowledge.

I am also very grateful to the Dean of SCSET and the Vice-Chancellor of Bennett University for their support and guidance. Their leadership has been instrumental in the success of this endeavor.

Thank you all for your invaluable contributions and support.

Contents

1

Geospatial Approaches for Environmental Pollution Mapping, Analysis and Mitigation

P. K. Garg

Introduction

Sustainable development refers to the human activities that try to balance the conflicting ideas of economic growth, maintaining the environmental quality and its viability (Skidmore et al. 1997, Brigham 2021). The concept of sustainable development mostly consists of three important parameters: environment, economy and society, as shown in Figure 1. It attempts to maintain a balance between the functions related to the productivity and conservation practices, and suggests optimal utilization of available natural resources based on their productivity, while keeping a good balance with the environment (Rao 2020).

Geospatial data may be defined as any data having a geographic component tagged with it (Garg 2022). It refers to real-world geographic objects of interest, such as rivers, buildings, lakes, countries, and locations. In addition to location, each object will also have attributes, such as a name, depth, or population. It can describe features, objects, phenomena or events that have a specific location—static or dynamic in nature. This type of data

Civil Engineering Department, Indian Institute of Technology, Roorkee, Roorkee-247667, Uttarakhand.
Email: pkgiitr@gmail.com

Figure 1. A typical model of sustainable development (Cafuta 2015).

is essential for many applications, such as vegetation planning, agriculture, disaster, environmental management, transportation planning, etc. (Bhan et al. 1997). Geospatial data can be collected in a variety of ways. Remote sensing satellites offer a large amount of geospatial data. Of late, vast amounts of information has also be gathered in a faster and effective way using unmanned aerial vehicles (UAVs)/drones (Garg 2019).

Geospatial approach terminology describes the variety of modern devices used to collect geospatial data as well as to map and analyze Earth's various features. It can be used for analysis, manipulations, modeling, simulations and decision making. Geospatial data enables mapping, monitoring, measuring, assessment, identification and management of resources. The geospatial approach can also be used to take crucial decisions based on available resources which are finite on the Earth. It is comparatively cheaper to use for a large number of environmental applications, particularly where the location is important.

Geospatial approaches include the use of Global Positioning Systems (GPS), remote sensing technology, LiDAR technology, Geographic Information Systems (GIS), and Cloud computing for geographic mapping and analysis of Earth (Garg 2022). These approaches are used to map and analyze the Earth's surface, to find real-time solution of problems, to interpret spatial patterns, and to forecast the future scenario. Geospatial approaches are utilized for a variety of environmental applications, such as traffic planning and management, biodiversity conservation, disaster management, forest fire monitoring and management, soil erosion modeling, vegetation monitoring, water pollution, air pollution, etc.

In the recent past, geospatial approaches have produced high resolution data and have been of help to decision-makers and planners. These have been used to understand the spatial patterns from large volumes of time series geospatial data. They are now considered primary driving technologies to analyze the spatial and temporal aspects of objects or phenomenon on the Earth. Geospatial approaches are used to generate smart maps that can be employed to make interactive query and get the desired results for policy-based decisions. Geospatial approaches have been found to be important tools for planning, management and decision making locally, nationally and globally (Garg 2022).

Geospatial Data

The whole world has been facing climate change and interconnected challenges. Geospatial data applied for environmental pollution mapping and monitoring has multiple benefits over the conventional approaches. The ever-increasing spatial, spectral and temporal resolutions of satellite data are considered to be the best not only for continuous monitoring and mapping of environmental parameters at local, regional and global levels, but also to help in making the appropriate decisions where environmental changes and their impacts are to be analyzed on a regular basis. Remotely sensed data along with other information collected from various sources and ground data can be used in several environment-related applications, as shown in Table 1.

Geospatial data can support organizations to understand and make quick decisions about real-world objects tagged to their locations (Garg 2020). This data (when large) may be utilized to visualize geographic patterns and relationships, analyze past trends, and predict future outcomes. Also, geospatial data integrated with other ancillary data, for example, demographic and socio-economic data, can be used to better visualise an area, which can help in effective decision-making.

Table 1. Environmental applications of remote sensing data (Ghauri and Zaidi 2011).

Atmospheric Parameters	Hydrological Parameters	Natural and Manmade Hazards	Land use Planning	Environmental Protection
– Aerosol – Fog – Black Carbon – Dust Storm – Ozone and other trace gases	– Water Quality – Soil Moisture – Sea Surface Temperature – Clouds Optical Properties – Snow Cover	– Floods, Tsunami, Earthquakes, Landslides Mapping and Risk Assessment – Droughts – Epidemic Mapping – Forest Fires	– Land use/Land Cover Changes – Urban Planning – Urban Heat Islands – Agriculture – Forests (Land & Coastal) – Coastal Zone monitoring	– Environmental Impact Assessment

In the 19th century, aerial photographs were used for map making. Light-weight cameras mounted on balloons and pigeons were also used. During the 20th century, aircrafts were used to collect aerial photographs. Satellite images were made available to the public in 1972 with the launch of an American satellite, Landsat. Its images were used to map and monitor the Earth's entire surface for the first time. Later, computers and software were introduced which allowed the storage and analysis of digital images, maps and data sets. Thereafter, GIS software were developed which were able to integrate a variety of geospatial data as layered maps and to carry out even more complex analysis. These layered maps are georeferenced together to get the real-location of any object on the ground, hence the term 'geospatial' was used. For more details of geospatial data, refer to Garg (2020).

Global positioning system (GPS) data

The GPS technology is used globally, today, for navigation and geolocation. It is a satellite-based navigation system, working in all weather conditions, anywhere globally, and 24 hours a day (Garg 2020). The GPS constellations make a network of 24 satellites which are launched by the US Department of Defense. GPS uses satellites that are orbiting the Earth to provide spatial and locational information. It has been fully operational since 1993. These satellites continuously send radio signals to GPS receivers, which determine the geographical position of any object using a combination of GPS satellites. GPS was originally devised for military applications, but in the 1980s, the US Government allowed the system to be used freely for civilian applications.

The working of GPS is based on the Trilateration principle. The technique requires a minimum of three satellites to accurately determine a particular 3D position on Earth. Tracking a minimum of four or more satellites or using differential GPS will give further higher positional accuracy. GPS has received great attention in the past due to its wider use in various applications. It has been used in resource mapping and management, as well as applications which require accurate geographic locations. Its integration with GIS provides an additional edge to determine the geographical characteristics of objects or phenomena. Today, all modern smartphones and smart watches also contain GPS.

Remote sensing data

Remote sensing involves mapping and monitoring of ground objects with the help of radiations that are reflected/emitted from various objects. These data may include aerial images from airplanes and drones; and visible, infrared, hyperspectral and microwave images from various satellites. Interpretation can be made by analyzing the data collected by the sensor platforms to determine the object's properties. Different kinds of remotely sensed satellite data with varying spatial and temporal resolutions are now available to assess and monitor phenomena on the Earth's surface. Remote sensing monitors the Earth systems, making it most suitable for addressing and solving global environmental, ecological, and socio-economic challenges (Avtar et al. 2020).

The first remote sensing satellite, Landsat 1, was launched in 1972, and since then remote sensing data has been successfully used to map and monitor natural resources and the environment. It provides the latest information on resources and a synoptic view of large areas regularly. Multispectral remotely sensed data provides additional information to analyze information on soils, water, land use/land cover, forests, etc. (Rao 2020). Remote sensing data has great advantages of providing synoptic view and repetitive coverage, and the capability of offering images of inaccessible areas at a relatively low-cost (Gibson 2000). Unmanned Aerial Vehicles (UAVs), also called drones, offer very high resolution data of the Earth surface from a lower altitude. Commercial satellites provide data at 1 m or smaller spatial resolution, and therefore these images are extremely useful for mapping and monitoring urbanisation, water, flood-loss, environmental pollution, etc.

The state-of-the-art remote sensing technology provides near real-time and precise information on several resources, such as land, water, forests, mineral resources that helps to map, monitor, manage and plan them (Garg 2019). Remotely sensed images offer a cost-effective solution to assess geospatial information, monitor land use and land cover changes, determine changes in landscape, check parameters that contribute to environmental pollution and distributions and monitor biodiversity.

Remote sensing image classification is a complex process, and requires careful selection of appropriate data (Lu and Weng 2007). As an example, Landsat/SPOT/IRS imagery can be used to monitor agriculture, urbanisation, water features, vegetation health; the IKONOS images for geospatial intelligence and to monitor urban infrastructure; the AVHRR (Advanced High-Resolution Radiometer) to study the impact of vegetation cover on global warming at global scale, the MODIS (Moderate Resolution Imaging Spectro-radiometer) Terra and Aqua sensors to monitor the atmospheric and oceanic composition in addition to the typical terrestrial applications. For more details of remote sensing satellites, refer to Garg (2022).

Advancements in remote sensing and free availability of large quantities of data from various satellites has given a fillip to the development of geospatial approaches, especially when combined with other data in GIS and AI and ML algorithms (Merghadi et al. 2020). The ML techniques, such as convolutional neural networks, random forests, and support vector machine, have been frequently utilised now to analyze remotely sensed data to assess and monitor environmental impact.

Point cloud data

Laser scanners collect geospatial data in a large number of tiny points of the object/terrain. Millions of these points collectively are called a point cloud. Each point has 3D coordinates (geographic location) associated, which when combined can be used to create a 3D model of an area/object. The point cloud data from LiDAR (Light Detection and Ranging) data using radio waves is a frequently used method for collecting spatial data at a very high resolution (Garg 2020). The LiDAR system is generally used to collect millions of points on the ground surface to create a DEM (Digital Elevation Model) or generate a digital twin of an object. These systems are available in a wide range from handheld to airborne LiDAR (Garg 2019). Table 2 presents the difference in characteristics of various remote sensing data collected from LiDAR, optical/mechanical and SAR.

The LiDAR uses light beams to measure distances, and is popularly known as laser scanning or 3D scanning. Based on the sensor used, LiDAR scanning devices can generate millions of pulses in one second that will hit the sighted object/surface and then bounce-off the objects/surface to finally reach to the LiDAR sensor. The returned pulsed laser measurements can be processed into a 3D visualization. The LiDAR-based maps can be used to get higher positional accuracy and to know how each object relates to other objects in terms of distance. The LiDAR point cloud data can be used to map entire city and to locate pollution sources in the area accurately. Objects, such as roads, bridges, ponds, polluting industry, tower, dam, buildings and trees, can easily be extracted.

Table 2. Characteristics of LiDAR, Optical/Multispectral and SAR data.

Parameters	LiDAR	Optical/Multispectral	SAR
Platform	Airborne system	Airborne/spaceborne system	Airborne/spaceborne system
Radiation	Has its own radiation	Uses reflected sunlight	Has its own radiation
Spectrum	Infrared beam	Visible/infrared beam	Microwaves beam
Frequency	Single frequency	Multi-frequency	Multi-frequency
Polarimetry	-	-	Polarimetric phase
Interferometry	-	-	Interferometric phase
Acquisition	Day and night	Only day time	Day and night
Cloud cover	Can't penetrate through clouds	Can't penetrate through clouds	Penetrate through clouds

Internet/open-source data

Web mapping is the technique of utilizing maps that are obtained by an information system for geospatial data. The maps available on the internet are generally static in nature, such as images (i.e., JPEG or TIFF) that may not give flexibility to users to assess the dynamic components. The internet-based maps are, however, easily available to share, can be customized, and made interactive with little effort. It would require technology/software for their successful implementation and drawing meaningful conclusions. The interactive maps have the obvious advantage that they can be easily customized to meet specific needs.

The availability of maps via Internet is a fast growing area. The Internet offers interactive and sharing capabilities which are the fastest approach to capture and analyze geospatial information. Web mapping software can analyze the online maps at the users' end. It may look fairly simple, but the techniques used for the analysis of these maps can be very technical and require expertise. In addition, software development is needed to support and analyze the web maps. Understanding the implementation of internet-based maps in GIS is crucial when working with the geospatial approaches.

Several software, such as Google Earth and web features, like Microsoft Virtual Earth have revolutionised the sharing and analyzing of geospatial data. The development in user interface is also gaining popularity amongst a wide range of users. Some of the software which can be used for webGIS activities is given below.

1. ArcGIS Online (AGOL) software is now mapping in the cloud service with several functionalities. It is suitable for large organizations as it is easy to integrate.

2. Carto (Postgres+PostGIS) web service can be used to control a large database. Although Carto is tailored to programmers, but some of its solution-based web mapping platforms require no development.

3. Mapbox is a programmer-centric like Carto. Mapbox is speedy, scalable, and can be used for customization. It is thus a preferred choice for high-traffic websites.

4. Mango Map uses an altogether a different approach. It provides the simplest approach to publish a web map. It provides the user interface with easy usability. Unlike Mapbox and Carto, it does not require any coding.

Geographic information systems (GIS)

A GIS consists of a computer program capable of storing, editing, manipulating, processing, and disseminating geospatial data and information as maps that are referenced to a geographical location. GIS software stores both spatial and non-spatial (attribute) data, and integrates them together to generate information and carry out the analysis (Garg 2020). A GIS can be visualised like a cake with many different layers, where each layer of a map presents a unique geographic theme, such as water bodies, land parcels, buildings, and forest. Each of these layers is georeferenced so that these are stacked on top of each other for further analysis.

Remote sensing offers most promising information which can be used as input data to a GIS software. It is a well-known fact now that several human activity is contributing to global warming, environmental changes, glacier melt, increase in glacial lake size, unpredicted non-seasonal rainfall, rapid change in land use and land cover, frequent forest fire, uncontrollable floods, massive landslides, and loss of agricultural crops. These problems need timely monitoring and management which can be supplemented by using geospatial data and approaches. Monitoring and mapping of the environmental parameters requires continuous detailed information that can be extracted using various geospatial data and approaches. GIS can prove to be an effective tool for environmental data analysis and to generate action plans. It offers better visualisation and makes it possible to study physical features and their relationships that may affect the environment of an area. For example, several factors, such as steepness of slopes, aspects, and vegetation, can be overlaid together to determine various environmental parameters and study their impact.

GIS can be used to analyze a comparative view of hazards (highly prone areas), risks (areas likely to have high risk), and safer areas. It can be used to effectively plan and manage environmental hazards and risks in a given region. To monitor environmental problems and devise action plans, the assessment of hazards and risks is most essential to plan mitigation activities. GIS can not only be used to provide support to activities related with environmental assessment, monitoring, and mitigation, but can also be used to develop environmental models. It can also support in hazard mitigation and planning, air pollution control, disaster management, forest fire management, natural resources management, wastewater management, detection of oil spills, and much more.

GIS comprises of a well-managed database to efficiently store, retrieve and manage geospatial data over large regions. It can be used to extrapolate the observations in case of scanty observations. It supports the statistical analysis of spatial distributions. With this unique capability, GIS can reveal patterns, relationships, and trends from large (big) geospatial data, required for making smarter decisions (Rao 2020). GIS software can be used to share ideas globally as to how to meet growing resource needs, scientifically plan land use, and safeguard the environment for future generations (ESRI

2007). It is an effective tool to understand spatial data and its relationship to people (urban sprawl), environment (forest cover data), or combined together (deforestation).

There are many commercial as well as open-source GIS software available. Open-source software, such as QGIS, and OpenStreetMap, are freely distributed and maintained by the open-source community. Easy availability of open-source software (e.g., SAGA, QGIS, SNAP, SeaDAS) has rapidly increased the number of users who make use of GIS technology to collect geospatial data to monitor natural resources and environment (Avtar et al. 2020). As a large number of inputs to GIS come from the remote sensing images, many GIS software are also used to work with remotely sensed data and to carry out several analyses with them. On the other hand, several GIS functions are also available in remote sensing software. Although many GIS commercial vendors and software companies, such as ESRI, Microsoft, Google and Intergraph, have contributed immensely in the past for development of GIS activities, the open-source software have gained popularity in the recent past in the GIS sector. Some of the available open source remote sensing and GIS software is given in Table 3. More details can be found in Garg (2020).

Geospatial AI

Geospatial artificial intelligence (GeoAI) is the amalgamation of GIS and modern technologies, for instance, AI (Artificial Intelligence), ML (Machine Learning), and DL (Deep Learning). The application of GeoAI may predict the future scenario or make projections. For example, in the healthcare industry, GeoAI maps our environment, linking the places we spend our time to environmental, social, and other factors that could potentially affect our health. GeoAI also analyses the dynamic geospatial data during natural disasters, examining damage levels and directing the relief efforts.

The ML and AI play a significant role in the analysis of geospatial data that enable development of complex models/algorithms and are subsequently applied to large datasets. The usage of 5G network as well as IoT (Internet of Things) devices offer enormous new data for effective and efficient geospatial analysis. In addition, acquisition of data from connected devices and sensors is near real-time for faster decision making. Cloud computing further allows processing and storage of huge volume of geospatial data in a compact and scalable way. It improves the accessibility and interoperability of geospatial data analysis. In addition, the use of VR (Virtual Reality) and AR (Augmented Reality) give new opportunity for visualization of geospatial data and its integration. These technologies

Table 3. Available open-source remote sensing and GIS software.

S.No.	Software	Details
1	Capaware	A fast C++ 3D GIS framework with a multiple plugin architecture for geographic analysis and visualization.
2	FalconView	A mapping system created by the Georgia Tech Research Institute for the Windows family of operating systems.
3	GDAL	A multiplatform set of tools for translating between geospatial data formats. It can also do reprojection and a number of geoprocessing functions. GDAL is designed for many applications both FOSS and commercial, including GRASS and QGIS.
4	GRASS GIS	A complete GIS software; originally developed by the US Army Corps of Engineers.
5	gvSIG	It is written in Java with several functionalities.
6	ILWIS (Integrated Land and Water Information System)	It easily integrates image, vector and thematic data.
7	JUMP GIS/OpenJUMP (Java Unified Mapping Platform	It is a desktop GIS OpenJUMP, SkyJUMP, deeJUMP and Kosmo emerged from JUMP.
8	Kalypso	It is supported by Java, GML3, and focuses mainly on numerical simulations in water management.
9	MapWindow GIS	It is a GIS desktop application and programming component.
10	OpebStreetMap	It is Google Maps alternatives. It has GPS navigation capabilities. It is editable. It is written in JavaScript with all the documentation regarding development and deployment.
11	OSSIM (Open Source Software Image Map)	A high-performance software compatible with over 100 raster and vector formats and more than 4000 types of projections.
12	PolSARPro	It has a wide range of tools, like radar decompositions, InSAR processing, and calibration. The functionality is similar to ArcGIS ModelBuilder, and it is easy to set up.
13	QGIS (Quantum GIS)	It runs on Linux, Unix, Mac OS X, and Windows.
14	SAGA (System for Automated Geoscientific Analysis) GIS	A hybrid GIS software which has a unique Application Programming Interface (API). It has a fast growing set of geoscientific methods, bundled in exchangeable Module libraries.
15	SeaDAS	It can be used for processing, display, analysis, and quality control of ocean color data. While the primary focus of SeaDAS is ocean color data, it is applicable to many satellite-based data analyses.
16	SNAP	SNAP and the individual Sentinel Toolboxes also support several sensors other than the Sentinel.

Table 3 contd. ...

...Table 3 contd.

S.No.	Software	Details
17	TerraView	It is a desktop software that handles vector and raster data stored in a relational or geo-relational database, i.e., a frontend for TerraLib
18	uDig	It is a desktop application (API and source code (Java) available)
19	Whitebox GAT	It is a useful GIS software with several functionalities

allow stakeholders to explore and understand the parameters contributing the environmental pollution in new ways.

Environmental Applications of Geospatial Approaches

The environment is the condition surrounding the areas where all things live. Various human activities contribute to land, water and air pollution in the environment. Because of unplanned resource utilization, many threats consisting of environmental destruction, contamination, global warming, scarcity and water shortage are being faced today. Environmental control is a useful action to manage and enhance the condition of environmental resources influenced by human actions. Geospatial data can be used to capture many environmental phenomena. In this digital age, geospatial approaches are providing enormous opportunities for developing new applications (Garg 2020). For instance, remote sensing images can be used to determine the spread of forest fires, the rate of increase of ocean water level, coastline changes, weather tracking (hurricanes or flooding), volcanic eruptions, etc.

Environmental monitoring relates to various activities to study and understand the current and past status of natural resources. Environmental data collected using geospatial data and approaches can be used for proper planning of conservation measures. It is used not only to monitor the resources, but also to protect them and mitigate the impact. Geospatial data greatly help in providing the relevant information required for Environmental Impact Assessment (EIA). By monitoring the changes in land cover, water quality, and vegetation health before and after various activities of a big project, geospatial data provides reliable information on environmental impacts of human activities which can guide formulating sustainable development practices and policies (Somvanshi 2021).

Several human activities require mapping and monitoring of environmental pollution. Examples include mining and manufacturing activities, civil construction projects (highways, railroads and electrical transmission lines), utility services, agricultural activities, etc. Remotely sensed data has great advantages for supporting the development of environmental pollution monitoring and related applications. It is a powerful tool for environmental studies, providing unique information that help understanding of our Earth environment.

The UAV/drone data are also an excellent tool for environmental pollution monitoring applications. The UAV/drone is especially useful in hard-to-reach areas. Drones equipped with RGB cameras, temperature sensors, humidity meter, pressure sensors, wind gauges, and various other sensors, are used to collect huge environmental data in greater details and at high resolution. For example, hazardous chemical spill which is rapidly moving a nearby river can be monitored with UAV data, and nearby industries could be alerted to toxic. Drones are efficiently used to monitor the environmental disasters in unsafe and hazardous locations, such as during the floods or after the landslides.

Real-time management of resources can be achieved through the use of technologies, like automated data collection and mapping, big data analytics and detection of land use and land cover changes. Adaption of geospatial technology is the best option that can provide a holistic approach to understand the complex interactions between the Earth's biophysical and social parameters in order to achieve a balance between developmental and environmental goals. Some applications of geospatial technology related to environment are discussed below (Garg 2019, Garg 2020, Garg 2022).

Monitoring of vegetation

In several environmental related studies, accurate and latest information about land use and land cover is important. Remotely sensed data provides very high-resolution images for the classification and mapping of different land use and land cover types. Information on forested areas and vegetation cover may be needed for urban planning and land use management. Geospatial data can be used to analyze vegetation cover, growth patterns, and its health. These data can also be used to assess changes in vegetation cover due to low rainfall, pest disease, or poor fertilizer, which may be important for resource allocation, agricultural planning, and biodiversity conservation.

Vegetation indices are effective and empirical indicators may be used to understand the condition of vegetation and the ecological conditions. In the past, a large number spectral indices have been developed using remotely sensed data to accurately map and monitor the resources and environment. These spectral indices are very effective for the interpretation and analysis of the processes related to change in environmental conditions, floods, soil moisture, and management of resources (Ban et al. 2017). Table 4 presents the major spectral indices that have been developed in the recent past from various remote sensing data types, and are applied in the field of environment mapping and monitoring.

Table 4. Major remote sensing based Spectral Indices developed for applications in environment mapping and Monitoring (Compiled from Avtar et al. 2020).

Features	Indices
Water	Normalized Difference Water Index (NDWI)
	Modified Normalized Difference Water Index (MNDWI)
	Automated Water Extraction Index (AWEI)
	Water Index (WI)
Chlorophyll content	Chlorophyll Index (CI)
Agriculture	Crop Water Stress Index (CWSI)
Vegetation	Ratio Vegetation Index (RVI)
	Leaf Area Index (LAI)
	Difference Vegetation Index (DVI)
	Normalized Difference Index (NDI)
	Perpendicular Vegetation Index (PVI)
	Moisture Stress Index (MSI)
	Physiological Reflectance Index (PRI)
	Photochemical Reflectance Index (PRI)
	Global Environment Monitoring Index (GEMI)
	Green Atmospherically Resistant Vegetation Index (GARI)
	Green-Normalized Difference Vegetation Index (GNDVI)
	Modified Simple Ratio (MSR)
	Soil and Atmosphere Resistant Vegetation Index (SARVI2)
	Modified Chlorophyll Absorption in Reflectance Index (MCARI)
	Specific Leaf Area Vegetation Index (SLAW)
	Transformed Difference Vegetation Index (TDVI)
	Wide Dynamic Range Vegetation Index (WDRVI)
	Enhanced Vegetation Index-2 (EVI2)
	Triangular Chlorophyll Index (TCI)
	Anthocyanin Reflectance Index (ARI)
	Canopy Chlorophyll Content Index (CCCI)
	Green-Red NDVI (GRNDVI)
Vegetation moisture	Aerosol Free Vegetation Index (AFRI)
	Global Vegetation Moisture Index (GVMI)
Vegetation and Forestry	Radar Vegetation Index (RVI)
Agriculture	Chlorophyll Absorption Ratio Index (CARI)
Vegetation and Agriculture	Atmospherically Resistant Vegetation Index (ARVI)
Vegetation and Forestry	Enhanced Vegetation Index (EVI)

Table 4 contd. ...

...Table 4 contd.

Features	Indices
Vegetation, Forestry and Agriculture	Normalized Difference Vegetation Index (NDVI)
Soil moisture	Shortwave Infrared Water Stress Index (SIWSI)
	Normalized Difference Moisture Index (NDMI)
Soil and Vegetation	Transformed Soil Adjusted Vegetation Index (TSAVI)
	Modified Soil Adjusted Vegetation Index (MSAVI)
	Optimized Soil-Adjusted Vegetation Indices (OSAVI)
Soil, Water and Vegetation	Tasseled Cap Transformation (TCT)
	Normalized Multi-band Drought Index (NMDI)
Soil, Agriculture, Vegetation, and Forestry	Soil-Adjusted Vegetation Index (SAVI)
Snow	Normalized Difference Moisture Index (NDSI)
Built-up area	Normalized Difference Built-up Index
Built-up and Bare land	Enhanced Built-Up and Bareness Index (EBBI)

Different remote sensing datasets, e.g., Landsat TM & OLI, SPOT, IRS and MODIS, have varying spatial resolution, spectral range and resolution, which may lead to different levels of mapping accuracy. With the fast development of spatial data, it is important that an appropriate dataset is selected. For example, MODIS vegetation index products are integrated with meteorological data to assess the spatial and temporal changes in regional vegetation.

Geospatial data can play a key role in biodiversity conservation. It provides precise information on habitats, species distributions, and ecosystem health, which can be used to locate areas of high biodiversity, detect the threats, and monitor effective implementation of conservation strategies.

Monitoring of mining areas

The over-exploitation of mineral resources is harmful for the surrounding environment. Such area may also experience soil erosion and land degradation. In the process of open-pit mining, mine areas encompass forests, grasslands, and farmlands and more. Geospatial data can regularly monitor the vegetation in mining areas, and provide a scientific approach for restoration of the environment. Remote sensing image can be utilised to monitor the ecological changes and its negative impacts on the mining area during the mining process.

Geospatial data is crucial in soil mapping and assessing soil erosion. Using spectral signatures curves, sensors operating in different wavelengths can be used to identify different soil types, soil moisture levels, and landscape changes due to soil erosion. This information is of prime importance for agriculture, land management and conservation planning.

Monitoring of protected areas

In protected areas, biodiversity, natural and cultural resources are protected and preserved by law or other means. Such areas may include nature reserves, biodiversity conservation areas, ecological areas, scenic locations, national forest parks, wildlife areas, birds' sanctuary, and natural and cultural heritage sites. These areas with their special entities are meant to protect important ecosystems, save endangered species, and protect natural historical heritages. They play a vital role in maintaining ecological balance. Geospatial data has been used to monitor land use and land cover change, temperature change, soil moisture, etc. The ecological and environmental changes in nature reserves may be correlated with human activities.

Air quality monitoring

Air quality has become an important public health issue world-wide. Remote sensing satellites, like NASA's Aura, can be employed to assess and monitor the air quality. This is done by measuring the concentration of pollutants, such as aerosols, carbon monoxide, and ozone. These data can be used to provide air quality forecasts, guide policy-making, and study the impact of air pollution on health. Various sensors and IoT system deployed for air quality monitoring, analytics, and planning, can accurately predict pollution levels. In addition, they show the geographic and temporal distribution of pollution within a region. They provide information about the areas which are most hazardous, and specifically important to be avoided by asthma patients. Drones equipped with thermal cameras can help quickly diagnose air leakages to improve a buildings' energy efficiency, as well as map major pollution sources in the city and their pathways in the rivers and other water bodies. GIS systems may be used to visualize and design planning activities. This, in turn, would help planners to take corrective actions to enhance air quality. Citizen participation is also an important consideration for such applications. For example, using the mobile app and geospatial location, they can alert the authorities about the areas that need immediate attention.

Climate change studies

Geospatial data plays a key role in understanding the climate change and its impacts. Satellite data can be used to monitor the sea level rise, snowmelt, atmospheric CO_2 and NO_x levels, and temperature trends, globally. The results can be used to develop predictive models and develop national and international policies to mitigate the climate change and its impacts.

Land surface temperature (LST)

Land surface temperature (LST) parameter is used in many studies. For example, it is required to study the exchange of matter and energy between land surface and atmosphere, global ocean circulation, climate change anomalies and so on. Neural network algorithms are now widely used in image processing, which use information from appropriate waveband as learning samples, and apply the deep learning neural network along with the radiation transfer model to retrieve the LST.

Water resources management

Geospatial data can be used to monitor and manage water resources, including lakes, rivers, canals, reservoirs, and coastal areas. Remotely sensed images can be used to assess the quantity and quality of water, detect pollution sources, detect changes in water levels, and water resource planning and management.

Coastal zones are very dynamic and vulnerable environments. Geospatial data permits continuous monitoring of these areas, providing precise information on parameters, like sea surface temperature, ocean color, and coastal landforms. This information is vital in managing fisheries, protecting marine habitats, and planning coastal defenses.

Disaster management

For disaster management, speed and accuracy of timely information is very crucial. Geospatial data provides real-time data on natural disasters, such as floods, wildfires, landslides, earthquakes, hurricanes and storms (Ritchie et al. 2022). It can be used effectively for prediction, monitoring, and mitigation of impacts of these events, which is essentially required for efficient post-disaster recovery and planning. These disasters pose threat not only to human-beings, but cause damage to property, infrastructure and environment.

Geospatial technology can be deployed for disaster risk assessment, simulation, and visualization. It can be used to guide during emergency responses, shelter operations, and post-disaster restoration and monitoring. Satellite images are also used in meteorology, weather forecasts, identifying the earthquake epicentres and dangerous locations. Many disasters need geospatial technology for proper management. GIS allows the integration of spatial and non-spatial information, and provides detail information which can be used in disaster management phases.

Wildfire management

Wildfire causes huge loss to flora and fauna. Although, wildfires are part of the natural cycle in a forest ecosystem, but when they create damages, their proper protection is essential. Geospatial data can be combined with other information, such as wind speed and direction, temperature, to make future predictions of fire spread, and plan the protection measures. GIS has proved to be very effective in forest fire management, including creation of fire hazard map, fire simulation, and forest resource management. For example, GIS can utilise several information layers, such as DEM and wind speed along with different models for forest fire management.

Healthcare

Geospatial technologies are important for global health activities. Public health surveillance and associated support systems require an exhaustive and precise geospatial base (Garg et al. 2022). Mobile phone data and big data analytics have already established their presence for COVID-19 monitoring and planning mitigation. Geospatial technology has been successfully used to track this disease around the world. For example, geospatial solutions based on Earth observations, sensors networking, and mobile contact-tracing coupled with AI, ML, and computer vision are being adopted very fast, and dominate the analysis of COVID-19.

The GIS has also been used to monitor other diseases, such as Malaria, HIV, etc. Today, all governments make use of geospatial information and databases on areas of disaster risks, including disease outbreaks and adverse weather patterns.

Summary

Geospatial data has information tied to a specific location on the Earth's surface, often represented as locational 3D coordinates. The scope of use of geospatial data is vast, as it covers every application where geographical position matters, such as ecology and environment, tourism, marine, agriculture, forestry, marketing, military, aircraft, law enforcement, logistics and transportation, demography, healthcare, meteorology, and many others. While most developed countries have made significant contribution to this area, many under-developed countries still lag far behind. High-quality and updated geospatial information is key to improve the environment, and is often overlooked by policy-makers. Reliable geospatial data enables policy makers, international organizations, civil society, and others to map and monitor the environment in an effective manner, and to devise appropriate policies. With increase of data collection and communication through smartphones, the geospatial approaches have now reached to grass-root level. Amalgamation of geospatial information into a modern system would be an asset for quick decision-making by governments and individuals.

Future trends and new research areas in geospatial data analytics may include use of open source data, geospatial data science, and predictive analytics using AI, ML and DL. Such advancements will not only improve the data quality, enhance computational efficiency, improve visualization and communication, foster interdisciplinary research, but also provide better tools and information for effective environment protection.

References

Avtar, R., Komolafe, A. A., Kouser, A., Singh, D., Yunus, A. P., Dou, J. et al. 2020. Assessing sustainable development prospects through remote sensing: A review. Remote Sens. Appl. 2020 Nov; 20: 100402. doi: 10.1016/j.rsase.2020.100402.

Ban, H.-J., Kwon, Y.-J., Shin, H., Ryu, H.-S. and Hong, S. 2017. Flood monitoring using satellite-based RGB composite imagery and refractive index retrieval in visible and near-infrared bands. Rem. Sens. 9(4): 313. https://doi.org/10.3390/rs9040313.

Bhan, S. K., Saha, S. K., Pande, L. M. and Prasad, J. 1997. Use of Remote Sensing and GIS Technology in Sustainable Agricultural Management and Development Indian Experience, Land Evaluation Congress: Geoinformation for Sustainable Land Management (SLM), 17 to 21 August 1997, ITC, Enschede, The Netherlands.

Brigham, Charles. 2021. UN Sustainable Development Goals Put a Global Spotlight on Local Action, ESRI, California, https://www.esri.com/about/newsroom/blog/sustainable-development-local-action/.

Cafuta, Melita Rozman. 2015. Open space evaluation methodology and three dimensional evaluation model as a base for sustainable development tracking. Sustainability 7(10): 13690–13712. https://doi.org/10.3390/su71013690.

ESRI. 2007. GIS Best Practices- GIS for Sustainable Development, ESRI, California, December 2007.

Garg, P. K. 2019. Introduction to Unmanned Aerial Vehicles, New Age International Pvt Ltd, Delhi.

Garg, P. K. 2020. Digital Land Surveying and Mapping, New Age International Pvt Ltd, Delhi.

Garg, P. K. 2022. Remote Sensing: Theory and Its Applications, New Age International Pvt Ltd, Delhi.

Garg, P. K., Tripathi, Nitin, Kappas, Martin and Gaur, Loveleen (eds.). 2022. Geospatial Data Science in Healthcare for Society 5.0, Springer Nature, Singapore.

Ghauri, B. M. and Zaidi, Arjumand. 2011. Application of Remote Sensing in Environmental Studies. Second International. Conference. On. Aerospace Science & Engineering At: Islamabad, Pakistan, December 2011.

Gibson, P. J. 2000. Introductory Remote Sensing. Principles and Concepts. London: Routledge.

Lu, D. and Weng, Q. 2007. A survey of image classification methods and techniques for improving classification performance. International Journal of Remote Sensing 28(5): 823–870. https://doi.org/10.1080/01431160600746456.

Rao, D. P. 2020. Role of Remote Sensing and Geographic Information System in Sustainable Development, International Archives of Photogrammetry and Remote Sensing, Vol. XXXIII, Part B7, Amsterdam.

Ritchie, Hannah, Rosado, Pablo and Roser, Max. 2022. Natural Disasters, https://ourworldindata.org/natural-disasters.

Skidmore, Andrew K., Bijker, Wietske, Schmidt, Karin and Kumar, Lalit Kumar. 1997. Use of remote sensing and GIS for sustainable land management. ITC Journal 1997-3/4.

Somvanshi Shivangi. 2021. Geospatial information to implement Sustainable Development Goals, Geospatial World. https://www.geospatialworld.net/blogs/geospatial-information-to-implement-sustainable-development-goals/.

2

AI-Driven Approaches for Pollution Mapping and Mitigation

Rashmi Kumari,[1,]* *Subhranil Das*[2] and
Raghwendra Kishore Singh[3]

Introduction

Air pollution encompasses all detrimental impacts arising from various sources which directly or indirectly contaminate the atmosphere or environment surrounding it (Ghorani-Azam et al. 2016). When air pollution initially became known as a threat to respiratory health, public health concerns grew to include a wide range of ailments, making it a serious threat to human health as an entire group (Almetwally and Ahmed 2020). World Health Organization (WHO) has released a paper stating that the risk presented by outdoor air pollution's small particulates is more than previously thought (Aayush et al. 2020). The WHO estimates that approximately 7 million people succumb to the effects of air pollution annually (WHO 2005). In both urban and rural regions, 5.2 million premature deaths worldwide were ascribed to outdoor air pollution in 2016 (mostly Particulate Matter) (WHO 2018). South and East

[1] School of Computer Science Engineering and Technology, Bennett University, Greater Noida, Uttar Pradesh, India – 201310.

[2] School of Business, Faculty of Business and Leadership, MIT World Peace University, Pune, Maharashtra, India – 411038.

[3] Department of Electronics and Communication Engineering, National Institute of Technology, Jamshedpur, India – 831014.

* Corresponding author: Rashmi.kumari@bennett.edu.in

Asia had the greatest mortality rates (15%) due to air pollution and its related dangers. Remarkably, between 1990 and 2017, the mortality rate increased by 14% in emerging countries like India, even while the number of deaths directly related to air pollution decreased (Mannucci et al. 2017).

Human health and life quality, influenced by various environmental factors such as psychosocial, physical, biological, and social elements, play a pivotal role in environmental health (Asha et al. 2022). Climate change, while having natural origin, has seen predominant human contribution since the 19th century, primarily through fossil fuel combustion, leading to CO_2 emissions. In a study, the regional and frequent shifts of CO_2 and particulate matter in Misurata, Libya, were carefully examined and provided informative data regarding the historical and regional aspects (Elsunousi et al. 2021). Cetin et al. (2019) measured quantities of CO_2 and particulate matter in different Bursa city regions. The findings revealed that CO_2 showed no statistical significance concerning seasons, while particulate matter exhibited a statistically significant 99.9% confidence level across seasons (Cetin et al. 2019). Another study focused on indoor CO_2 concentration during examinations, highlighting elevated levels exceeding 1500 ppm within the initial 10 minutes (Cetin et al. 2016). Additionally, an investigation into Pb and Cr pollution in Turkey's capital city, involving topsoil samples from 50 regions, emphasized the severe environmental and ecosystem threats posed by air pollution. The current accelerated warming of the Earth, unprecedented in recorded history, is disrupting nature's equilibrium, leading to intensified storms, floods, cold spells, and heatwaves with substantial socioeconomic repercussions. Heat waves (HWs) are predicted to become more often and intense due to human-caused climate change, which will quickly alter patterns of biodiversity and the structure and function of ecosystems (Hobday et al. 2016). In order to address these issues, reliable forecasting methods for weather, air pollution, and climate change must be developed, along with efficient early-warning systems.

Artificial Intelligence (AI) stands as an emerging technological discipline focused on exploring and advancing theories, methods, technologies, and application systems to simulate, enhance, and broaden human intelligence (Subhranil et al. 2021, Rashmi et al. 2023). Giving machines the ability to perform complex activities that usually need human comprehension is one of artificial intelligence's main goals. Among the notable benefits of AI are automated data extraction (Chaurasia et al. 2020a), improved labor and time efficiency, convenience, and long-term sustainability (Chowdhury et al. 2012). These merits facilitate diverse applications, spanning disease prediction (Inampudi et al. 2021, Ayachit et al. 2020), weather forecasting (Chaurasia et al. 2020b), the construction of expert systems, environmental monitoring, and the prediction of pollutants (Hashimoto et al. 2020). Artificial Intelligence

(AI) aims to mimic human decision-making and problem-solving abilities. Figure 2 depicts the connection between machine learning algorithms, deep learning algorithms, and AI strategies. To reduce the hazards of public exposure to environmental pollution and assist policymakers in creating more effective environmental protection regulations, AI must be used as a key tool for environmental pollution forecasting. AI is particularly good at anticipating and forecasting air pollution because it is skilled at managing complex and non-linear interactions among spatial-temporal parameters (Subhranil et al. 2021). It is excellent at monitoring the state of pollution right now and offers a quick and accurate way to locate regions that are hotspots for pollution. Extreme weather prediction is another use for deep learning, a subclass of artificial intelligence. AI's remarkable capability to manage vast amounts of data and tackle real-world complex problems enhances accuracy in weather forecasting and modeling, thereby offering valuable insights to policymakers and decision-maker. To address the aforementioned challenges effectively, it is crucial to implement early-warning systems by forecasting air pollution as shown in Figure 1.

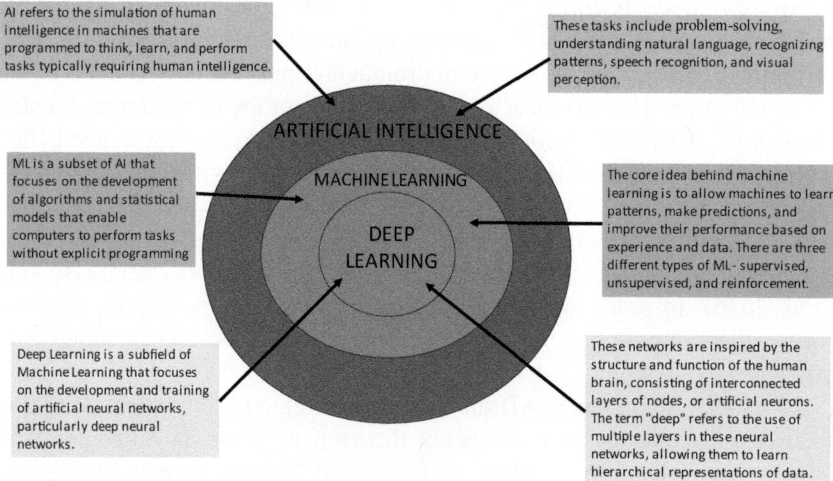

Figure 1. Overall view of AI, ML and Deep Learning.

Different Types of Pollutants

Air pollution is one of the critical global crises, wielding a direct and multifaceted impact on both environmental and human health. The repercussions extend beyond ecological concerns, manifesting in significant health crises marked by premature mortality, respiratory ailments, strokes, and cardiovascular diseases (Ong et al. 2016). This problem stems from a combination of natural and man-made events that create a hazardous particle mixture that leads to numerous more disasters. Natural events like sudden forest fires that produce dangerous pollutants add to the complexity of the issue (Bai 2018). The air we breathe is becoming more and more contaminated due to the increase in industrial and automotive emissions in modern times. Most air pollutants, which include CO_2, NO_2, SO_2, CO, and particulate matter, have a negative impact on both people and ecosystems. Figure 2 visually encapsulates these major pollutants, elucidating their causative roles in environmental degradation and human health impacts. The fallout includes exacerbating environmental problems like the greenhouse effect, ozone depletion, and photochemical smog, ushering in significant environmental disasters (Najjar et al. 2011). The toll on human health is further compounded by factors such as chronic diseases, with volcanic eruptions and industrial expansion emerging as major contributors of sulphur dioxide (SO_2). Beyond its role in health complications, the presence of SO_2 adds to environmental woes by potentially contributing to acid rain, a phenomenon that detrimentally affects ecosystems. The interconnectedness of these elements underscores the urgency of addressing air pollution, as the consequences ripple through human well-being.

The omnipresence of air pollution presents an undeniable reality, with its repercussions acknowledged to be more severe compared to soil and water pollution (Masood et al. 2021). Governments worldwide are actively pursuing strategies to enhance air quality, enacting diverse policies and initiatives. Fuel combustion in industries and transportation stands out as a major contributor to NO_2 emissions, causing adverse effects on human health, particularly

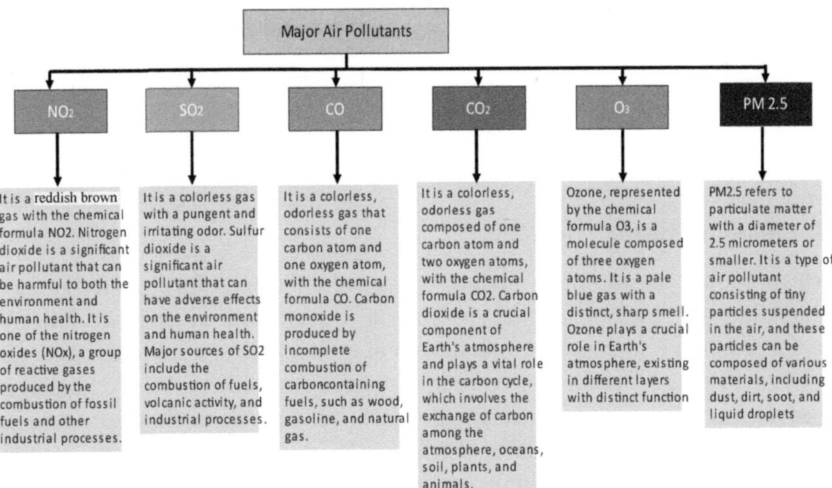

Figure 2. Overall summary of different types of pollutants in air pollution.

escalating the prevalence of lung diseases. Ground-level ozone, renowned for its potent oxidizing properties, poses severe health risks to humans upon prolonged exposure, significantly compromising pulmonary function and potentially leading to conditions like lung cancer. Particles of varying diameters, with those smaller than 1 μm capable of reaching the alveoli, contribute to respiratory issues and, in some cases, impair cognitive function and contribute to mental illnesses (Han et al. 2017).

Methodology

Various techniques are utilized to categorize air pollutants, as depicted in Figure 3. The forecasting of distinct air pollutants is addressed through four models: traditional statistical models, AI models, hybrid models, and 3D models. A detailed explanation of each model is provided in the subsequent subsection.

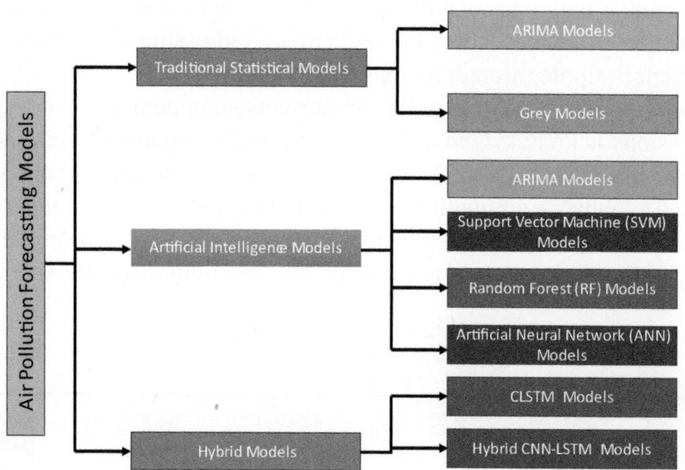

Figure 3. A summary of the models used to forecast air pollution.

Traditional statistical models

Air pollution forecasting models are essential tools for predicting and managing air quality, aiding in the mitigation of adverse health and environmental impacts. Because statistical models are simple to use and can estimate future pollutant concentrations by analyzing the association between past data's climatic parameters and pollutant concentration, they are useful in forecasting expected air quality. Importantly, these models operate without detailed knowledge of pollution sources. Traditional statistical models, such as ARIMA and the GM have been historically employed in this domain

(Arruda et al. 2023). These models operate by establishing connections between various air quality variables based on feasibility and statistical averages. However, their effectiveness in achieving high forecasting accuracy is often limited. The ARIMA model has been specifically employed to PM2.5 prediction because to its performance history. Nevertheless, challenges arise in capturing the regional characteristics inherent in air quality dynamics, which can be complex, chaotic, and nonlinear. Nonlinear fitting, in particular, presents a drawback for these traditional statistical models (Kaur et al. 2023).

To overcome the limitations of traditional models, researchers have explored alternative approaches. One such avenue involves leveraging advanced machine learning techniques and statistical methods. SVM, RF, LSTM, Generalized Additive Models (GAM), Kriging models, Artificial Neural Networks (ANN), and quantile regression are among the models investigated. These models offer diverse strategies to address the nonlinearity, spatial variability, temporal dependencies, and complex relationships inherent in air quality data. Support Vector Machines, for instance, excel in capturing nonlinear patterns, while Random Forest models handle complex interactions. Because long short-term memory models are good at capturing temporal relationships in time-series data, air pollution forecasting benefits greatly from their use. Generalized Additive Models provide flexibility in modeling complex relationships, and Kriging models are well-suited for spatial interpolation. Artificial Neural Networks, with their capacity for learning intricate patterns, have gained popularity, and quantile regression allows for a more comprehensive understanding of the uncertainty associated with air quality predictions.

Artificial Intelligence (AI) models

Several scholarly investigations confirm that in many applications, artificial intelligence (AI) models outperform statistical models as well as physical forecasting techniques (Chen et al. 2020). Artificial Intelligence (AI) includes machine learning, which is known for its efficiency in categorization and regression analysis. AI emulates human vision, learning, and reasoning. In the realm of pollutant forecasting, machine learning, particularly through techniques like BPNN, ELM, LSTM, WNN, and other models including SVM has demonstrated remarkable resilience and accuracy. Artificial neural networks, such as BPNN, emulate the nonlinear series modelling of the human brain. The advent of deep learning models, characterized by multiple layers, has further expanded the capabilities of forecasting algorithms. In the realm of air quality, research is still being done to determine the distinctive qualities of various intelligent prediction models, realizing that different techniques are needed to address different issues (Hashimoto et al. 2020).

ANN has proven notably adept at handling increasingly intricate tasks, and its application extends successfully to air pollution prediction through a sophisticated algorithm (Guo et al. 2020). By utilizing crucial data from regularly available meteorological factors, Elangasinghe et al. (2014) created an artificial neural network model for air pollution prediction that, under specific circumstances, successfully caught the temporal changes in pollutant concentrations. Pardo and Malpica (2017) choose to forecast Madrid's air quality using a dual-layered LSTM neural network. The utilization of deeper LSTM networks enhances predictive precision, albeit at the expense of heightened computational demands and time investment (Zeng et al. 2022). In a novel approach, Song et al. (2019) created a hybrid model that included Kalman filtering and LSTM to evaluate concentration of air quality components.

Hybrid models

In order to get improved performance, hybrid models combine several methods or approaches with the advantages of each component part (Song et al. 2019). Figure 3 illustrates various Hybrid AI models which are designed in forecasting pollutants.

Wen et al. introduced a novel spatiotemporal C-LSTME model (Wen et al. 2019). To improve forecasting results, the model integrates meteorological and particle data. However, due to the complexity of the weather and temporal feature model frameworks and their requirement for a high number of data records, the temporal complexity of the suggested solution is significantly increased. Another predictive model for PM2.5 concentration was developed by Zeng et al. (2022) combining the NLSTM neural network and the ESWT (Liu et al. 2021). This technique improves prediction performance by blending deep learning technology (Zeng et al. 2022). In a different study, air pollution was predicted using RFR, DTR, and LR techniques. The results showed that RFR performed better on the provided dataset than LR and DTR (Shivakumar et al. 2022). Althuwaynee et al. assessed the correlation clusters of PM10 and other contaminants using decision tree algorithms to assess the risk of air pollution (Althuwaynee et al. 2020). Shaziayani et al. employed a variety of tree-based machine learning techniques, including boosted regression trees, random forests, and decision tree algorithms, to predict the PM10 concentration for the upcoming day (Shaziayani et al. 2022). In order to increase prediction accuracy and time performance, Wang and Kong developed a model to forecast air quality using enhanced decision tree techniques (Wang et al. 2019). The usefulness and dependability of the GRNN technique for accurate PM2.5 level forecasting in urbanized areas were shown by Yan et al.'s application of GRNN to predict PM2.5 (Yan et al.

2021). Bekker et al. (2021) employed a spatial-temporal feature integration of meteorological data, historical pollutant records, and PM2.5 levels from nearby stations. They evaluated how different Deep Learning algorithms performed. According to the experimental results, their hybrid CNN-LSTM multivariate strategy outperformed standard models in predictive accuracy and produced more accurate predictions (Gilik et al. 2020). These hybrid AI techniques exhibit increased reliability and accuracy, particularly in the development of environmental pollution monitoring.

Prediction of Pollutants Based on AI Techniques

The sources of pollutants, including Particulate Matter (PM), Carbon Monoxide (CO), Carbon Dioxide (CO_2), Oxygen (O_2), Sulfur Dioxide (SO_2), and Nitrogen Dioxide (NO_2) are described below.

PM is a complex mixture of tiny particles and liquid droplets in the air. Primary sources include vehicle exhaust, industrial processes, construction activities, and agricultural practices. Secondary sources include chemical reactions in the atmosphere. Particularly the fine particles (PM2.5) that can enter the lungs deeply can have detrimental impacts on health. Cardiovascular and respiratory disorders are linked to it. The primary source of CO is incomplete combustion of fuels containing carbon. The emissions from automobiles, industrial operations, and home heating systems are common sources. In high amounts CO can be lethal and cause symptoms like headaches and dizziness by interfering with the body's ability to transfer oxygen.

Carbon dioxide (CO_2) is a naturally occurring element of the Earth's atmosphere, resulting from the burning of fossil fuels such as coal, oil, and natural gas, as well as deforestation and several industrial operations. Excess CO_2 adds to the greenhouse effect, which causes global warming and climate change even though it is essential for plant photosynthesis. The primary source of O_2 in the Earth's atmosphere is photosynthesis, which occurs in plants, algae, and cyanobacteria. Essential for aerobic respiration in animals and humans, O_2 supports life. Changes in oxygen levels can affect ecosystems and aquatic life. Burning fossil fuels like coal and oil that contain sulfur releases sulfur dioxide (SO_2). Significant sources are industrial processes, especially those involved in metal smelting. In addition to contributing to the development of acid rain, which damages soil, water bodies, and vegetation, SO_2 can lead to respiratory issues. NO_2 is produced by combustion processes in vehicles and industrial facilities. Agricultural activities and natural sources, like lightning, also contribute. NO_2 is a major component of air pollution, contributing to respiratory issues. It also contributes to the creation of particulate matter and ground-level ozone. Understanding and mitigating these pollutant sources are crucial for addressing air quality and promoting public and environmental health.

Prediction of ozone (O₃) concentration

Ozone, a potentially harmful gas due to its strong oxidizing properties, can have adverse effects on living organisms when present on the Earth's surface. At higher levels, it can cause irritation of the skin, eyes, and nose in addition to lesions in the upper respiratory tract. High ozone levels can also harm agricultural crops, and the loss of plantations could make the situation worse in the future. Murillo et al. developed a multi-pollutant prediction technique using ANN and SVR technology (Murillo-Escobar et al. 2019). Four climate variables and hourly pollutant concentrations were fed into the AI models, and the Particle Swarm Algorithm was used to minimize computational complexity. The efficiency of the SVR-PSO technique was shown by the results in all seasons. A dual-scale ensemble learning model was introduced by Zhou et al. (2019) to estimate daily ozone levels in heavily polluted Chinese cities. Their innovative model outperformed comparison models, as demonstrated by lower RMSE values. A fuzzy binary relation-based technique was employed by Chattopadhyay et al. (2021) to study air pollution. They found that NO_2 and O_3 were the main factors influencing the overall Air Quality Index. Alimissis et al. (2018) evaluated predictions of various pollutants, highlighting the superiority of ANN over MLR in terms of accuracy. Ozone Concentration has been shown in Figure 4.

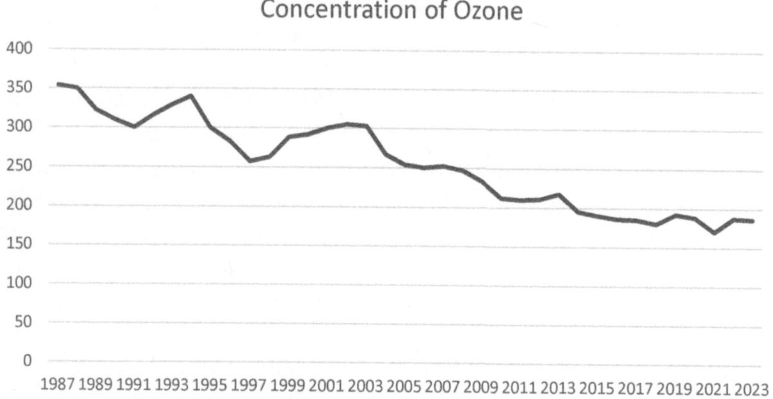

Figure 4. Ozone concentration during 1987 to 2022.

Prediction of Carbon Monoxide (CO) concentration

Haemoglobin within the human body reacts with carbon monoxide (CO), forming carboxyhaemoglobin (CO-Hb), which diminishes the blood's capacity to carry oxygen and can be fatal in high concentrations over a short duration. Particularly sensitive populations, such as infants, the elderly, or people with heart or lung problems, are most dangerous from carbon monoxide poisoning. Thus, forecasting atmospheric carbon monoxide concentrations is essential for putting into place efficient early-warning systems. Zhou et al. (2010) assessed the performance of a novel prediction model integrating SVM and partial least squares (PLS) for estimating CO concentrations, with the PLS-SVM ensemble technique proving superior. Slini et al. (2003) explored the uncertainty of CO predictions using SVR, ANN, and ANFIS, concluding that SVR models offer an acceptable range of uncertainty for CO forecasting. Mishra and Goyal (Mishra et al. 2015) used a neuro-fuzzy model that took into account a number of variables to precisely estimate CO levels in Delhi's urban environments. Chattopadhyay et al. (2021) studied air pollution over Kolkata, India, during the post-monsoon to winter transition using a fuzzy binary relation-based approach. The collective findings emphasize the uniqueness of each forecasting model in predicting CO, with AI-based models standing out for their superior accuracy and reliability in CO forecasting. Million deaths due to CO during the years of 2010 to 2022 as shown in Figure 5.

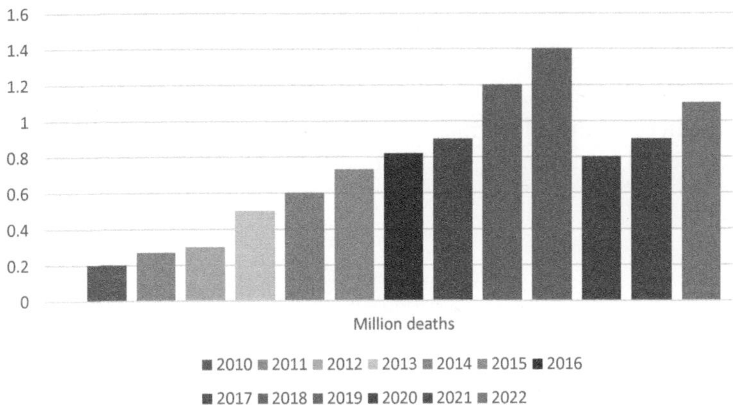

Figure 5. Million deaths due to CO during the years 2010 to 2022.

Prediction of CO_2 concentration

CO_2 is a significant greenhouse gas with heat-trapping characteristics, contributing to global warming. High concentrations of CO_2, especially above 5000 parts per million (ppm), can pose health risks, leading to respiratory and cardiovascular issues by lowering the pH of blood serum and causing acidosis. It's crucial to develop accurate models for forecasting CO_2 emissions, considering both environmental concerns and potential health impacts. Researchers have employed Artificial Neural Networks (ANN) in these predictive models, showcasing their effectiveness. Sohn et al. (1999) utilized ANN to predict various pollutants, demonstrating the versatility of this approach in handling multiple pollutants simultaneously. The accuracy of ANN in estimating carbon dioxide emissions was highlighted by Ahmadi et al. (2020), who concentrated on utilizing it to anticipate emissions in Middle Eastern nations. The ANFIS has also proven valuable in predicting CO_2 emissions. ANFIS was effectively used by Norhayati and Rashid (2018) to forecast emissions from a clinical waste incinerator facility, showcasing the adaptability of this model in real-world scenarios.

Additionally, in the context of Bahrain's CO_2 emissions, Qader et al. (2021) explored various strategies, including Holt's methods, with neural network models demonstrating superior performance. The neural network models demonstrated the lowest root mean square error (RMSE), demonstrating their efficacy in representing the intricacies of CO_2 emission patterns within the area. Additionally (Wei et al. 2017) presented a model that uses an improved extreme learning machine and factor analysis to estimate the intensity of carbon emissions. This further emphasizes the versatility of neural network-based approaches, specifically multi-layer perceptron ANN, in addressing challenges associated with time series forecasting and nonlinear relationships. These studies illustrate the robustness and reliability of neural network models, particularly ANN and ANFIS, in forecasting CO_2 emissions across various contexts. These models prove valuable tools in understanding and predicting the dynamics of CO_2 concentrations, contributing to environmental management and health protection as shown in Figure 6.

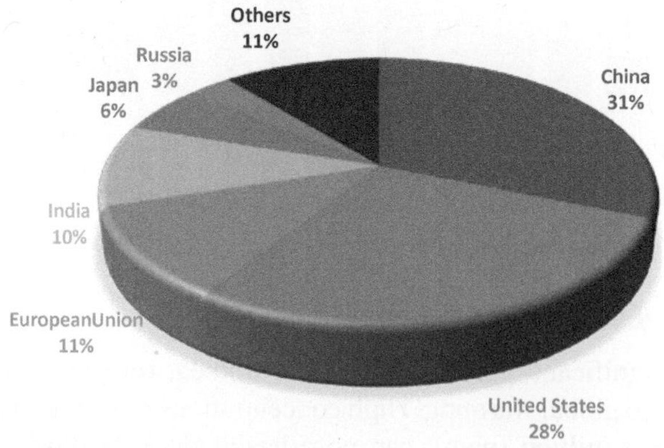

Figure 6. Percentage of emission of CO_2 gasses in different countries.

Prediction of Nitrous Oxide (NO$_X$) and Sulpur Dioxide (SO$_X$) concentration

Emission of nitrogen dioxide (NO$_2$), which is mostly caused by burning fossil fuels in transportation and industrial activities, is a major source of pollution in the environment and a health danger to ecosystems and people alike. NO$_2$ is a key contributor to acid rain, which can have detrimental effects on water, land, vegetation, buildings, and overall human health. The exposure to NO$_2$ has been linked to an increased prevalence of respiratory illnesses. Sulfur dioxide (SO$_2$) is another major air pollutant, and when combined with NO$_2$, it further intensifies the impact on acid rain. To address the complexities of these pollutants and their effects, researchers have focused on developing accurate models for predicting and warning about NO$_2$ and SO$_2$ concentrations. Various advanced techniques, particularly Artificial Neural Networks (ANN), have been employed in these models. Studies, such as the one by Slini et al. (2003), highlight the importance of hourly predictions for NO$_2$, CO, and ozone, emphasizing the need for accurate and dynamic forecasting models. Hybrid models, like those incorporating the Grey Wolf Optimizer, have been proposed by (Xu et al. 2017) to predict concentrations in specific regions of China. These models aim to improve the understanding of pollutant dynamics and enhance the precision of forecasts. PCA based ANN models and Multiple Linear Regression (MLR) models have also been explored for estimating NO$_2$ concentrations. Wang et al. (2018) introduced adaptive forecasting models that integrate ANN and SVM techniques, showcasing the potential for improved air quality predictions through hybrid approaches. Different neural network architectures, such as the Elman model and Multi-layer Perceptron (MLP), have been utilized to forecast concentrations of pollutants like SO$_2$. Shams et al. (2021) compared MLP and MLR for SO$_2$ emphasizes the advanced predictive capabilities of ANN models over traditional regression methods. Innovative two-step hybrid models, incorporating techniques like CEEMD, SVR, GWO, and Cuckoo Search (CS), have been proposed for SO$_2$ and NO$_2$ forecasting. These models, exemplified by Zhu et al.'s CEEMD-CS-GWO-SVR (Zhu 2019), demonstrate enhanced accuracy in predicting pollutant concentrations, providing valuable insights for air quality management. The novel approach of Liu et al. (2021), combining LSTM networks, showcases a sophisticated method for forecasting NO$_2$ concentrations. Their model structure, despite certain trade-offs in performance metrics, represents a promising step forward in refining predictions for this pollutant. These studies collectively underscore the importance of employing advanced modeling techniques, particularly neural network-based approaches, to gain a nuanced understanding of NO$_2$ and SO$_2$ dynamics, ultimately improving the accuracy of predictions for

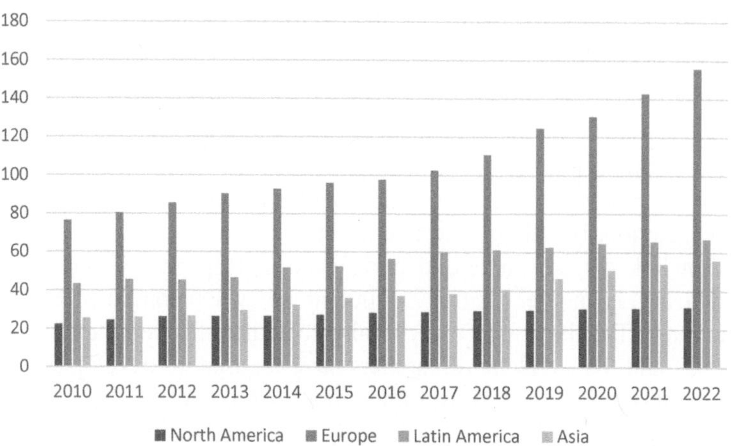

Figure 7. Market share (in Millon dollars) for supply of NO_2 in different countries.

effective air quality management and public health protection as shown in Figure 7.

Prediction of Particulate Matter (PM) concentration

Particulate Matter (PM) is a complex mixture of tiny particles suspended in the air, and its adverse health effects make accurate forecasting crucial. In recent years, researchers have explored various artificial intelligence (AI) based models to predict PM concentrations with greater precision and efficiency than traditional statistical methods. These advancements aim not only to provide accurate forecasts but also to enhance our understanding of the intricate relationships between various factors influencing air quality. Fernando et al. (2012) delved into PM10 concentration forecasting using a Multilayer Perceptron algorithm, a type of ANN. This approach yielded accurate results, as indicated by performance metrics such as Index of Agreement (IA), RMSE, R^2, and MAE. The study showcased the potential of neural network models in capturing the complex patterns inherent in PM concentration data. Cheng et al. (2019) introduced hybrid models, including wavelet-ANN, wavelet-SVM, and wavelet-ARIMA, with the goal of improving prediction capabilities. These models leveraged the strengths of wavelet transformations and different AI techniques, demonstrating enhanced performance compared to individual methods. The emphasis on hybridization reflected a growing trend in utilizing diverse approaches for more robust predictions. Mirzadeh et al. (2022) explored wavelet-support vector regression for PM10 concentration prediction. The study demonstrated that wavelet transform-related AI strategies are more effective than standalone AI models in terms of performance. This approach not only improved accuracy but also demonstrated the role of wavelet transforms in enhancing the predictive capabilities of AI techniques. Hu et al. (2017) introduced ensemble machine learning techniques, namely XGBoost-RF-ARIMA, highlighting the possibility of merging several models for PM2.5 concentration prediction. Ensemble methods, which integrate predictions from multiple models, have gained popularity for their ability to mitigate individual model weaknesses and improve overall accuracy. Zhu et al. (2019) introduced the improved CEEMD model, a hybrid forecasting approach for daily PM2.5 concentration. This model, integrating CEEMD, SVR, and GWO, demonstrated superior accuracy. The study highlighted the importance of selecting appropriate methods and optimizing their combination for effective forecasting. Feng et al. (2015) explored a unique hybrid technique for PM2.5 concentrations. This model, incorporating WPD, Bi-LSTM, SAE, and NSGA-II outperformed other benchmark models. The emphasis on combining different AI techniques showcased the potential benefits of multi-layered hybrid models. Bai et al. (2018) presented EEMD-LSTM, an ensemble empirical mode decomposition with long short-term memory, for hourly PM2.5 prediction. This model performed better on measures like Mean Absolute Percentage Error (MAPE), correlation coefficient, and root mean square error (RMSE) than feed-forward neural networks and LSTMs. The study underlined how crucial model selection is to increase prediction accuracy. For daily PM2.5 concentration estimates, Sun et al. (2017) developed a method based on PCA and LSSVM supplemented by cuckoo search. The results demonstrated the effectiveness of this approach in outperforming general regression neural network methods and single LSSVM models. The study showcased the significance of feature extraction and optimization techniques for accurate PM2.5 concentration predictions. Liu et al. (2021) proposed WPD-PSO-BP-Adaboost, a hybrid model for PM2.5 concentrations. This model, combining wavelet packet decomposition (WPD), particle swarm optimization (PSO), backpropagation neural networks (BP), and AdaBoost, demonstrated enhanced forecasting precision in multi-step predictions. The study highlighted the benefits of incorporating diverse techniques to address different aspects of PM concentration forecasting.

These studies underscore the versatility and effectiveness of AI-based hybrid models in predicting PM concentrations. The exploration of various techniques, including neural networks, ensemble methods, and optimization algorithms, reflects a comprehensive approach to addressing the challenges posed by PM forecasting. As researchers continue to refine and innovate these models, the potential for more accurate and reliable air quality predictions becomes increasingly promising. The emphasis on precision, nonlinearity, and the ability to outperform traditional statistical models underscores the transformative impact of AI-driven approaches on the field of air quality forecasting as depicted in Figure 8.

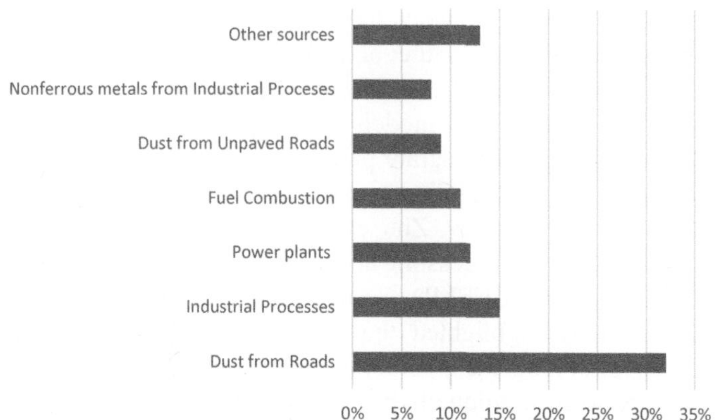

Figure 8. Distribution of percentage of different sources of Particulate Matter.

Discussion

In the expansive realm of healthcare, artificial intelligence has notched remarkable achievements, particularly in revolutionizing disease detection within the healthcare system. Its significant strides are evident in the precise identification of diseases such as cancer and infections, ushering in a new era of diagnostic capabilities. Furthermore, AI has been essential to the development of surgical robots, which has enhanced surgical techniques and increased diagnostic precision. The applications of AI extend across various healthcare domains, encompassing early sickness detection, cancer identification, management of chronic conditions, drug development, advancements in robot-assisted surgeries, enhancements in healthcare delivery, and facilitating scientific research. A notable facet of this technological progression is the emergence of wearable technologies and applications that leverage AI methodologies. These innovations analyze vital signs to forecast potential health issues, empowering individuals with proactive health management tools. Going deeper into the topic, this section focuses on how AI and machine learning are being used to forecast long-term respiratory problems. It examines the complex relationship between air pollution and human health, providing information on how air pollution affects COVID-19 infection and death. The narrative then pivots to underscore the pivotal role of AI in the healthcare system, with a specific focus on diagnosing respiratory problems. The discussion encompasses the predictive aspects of AI in managing cardiopulmonary diseases, showcasing its potential to revolutionize healthcare delivery. Furthermore, the section sheds light on how AI and ML methodologies can be harnessed to monitor and control acute exacerbations. This multifaceted exploration aims to underscore the far-reaching impact of air pollution on human health while concurrently emphasizing the transformative role played by AI technologies in shaping the landscape of modern healthcare. In essence, it provides a comprehensive overview of the intricate dynamics involving air pollution, human health, and the innovative integration of artificial intelligence within the healthcare paradigm.

Conclusion

In conclusion, this paper provides a thorough examination of the role of AI in addressing environmental contamination and promoting sustainable resource management. The review highlights the benefits, features, and limitations of various AI techniques, emphasizing their potential in pollution control and mitigation. Expert systems, fuzzy logic, and neural networks stand out as effective tools for process control in pollution management. The study underscores the transformative impact of AI in analyzing intricate environmental data, optimizing resource utilization, and predicting pollution trends. As society grapples with urgent environmental challenges, the integration of AI emerges as a powerful ally, offering the capacity to enhance sustainability practices. By harnessing AI-driven technologies, we can actively combat environmental degradation while ensuring efficient and responsible use of natural resources. The findings underscore the pivotal role of AI in steering us towards a more environmentally conscious and sustainable future.

References

Aayush, K., Vishal, D., Hammad, N. and Manu, K. S. 2020. Application of artificial intelligence in curbing air pollution: The case of India. Asian Journal of Management 11(3): 285–290.

Ahmadi, A., Fadaei, Y., Shirani, M. and Rahmani, F. 2020. Modeling and forecasting trend of COVID-19 epidemic in Iran until May 13, 2020. Medical Journal of the Islamic Republic of Iran 34: 27.

Alimissis, Anastasios, Kostas Philippopoulos, Tzanis, C. G. and Despina Deligiorgi. 2018. Spatial estimation of urban air pollution with the use of artificial neural network models. Atmospheric Environment 191: 205–213.

Almetwally, Alsaid Ahmed, May Bin-Jumah and Ahmed A. Allam. 2020. Ambient air pollution and its influence on human health and welfare: an overview. Environmental Science and Pollution Research 27: 24815–24830.

Althuwaynee, Omar F., Abdul-Lateef Balogun and Wesam Al Madhoun. 2020. Air pollution hazard assessment using decision tree algorithms and bivariate probability cluster polar function: evaluating inter-correlation clusters of PM10 and other air pollutants. GIScience & Remote Sensing 57(2): 207–226.

Arruda, Helder Moreira, Rodrigo Simon Bavaresco, Rafael Kunst, Elvis Fernandes Bugs, Giovani Cheuiche Pesenti and Jorge Luis Victória Barbosa. 2023. Data science methods and tools for Industry 4.0: a systematic literature review and taxonomy. Sensors 23(11): 5010.

Asha, P., Natrayan, L. B. T. J. R. R. G. S., Geetha, B. T., Rene Beulah, J., Sumathy, R., Varalakshmi, G. et al. 2022. IoT enabled environmental toxicology for air pollution monitoring using AI techniques. Environmental Research 205: 112574.

Ayachit, S. S., Kumar, T., Deshpande, S., Sharma, N., Chaurasia, K. and Dixit, M. 2020, December. Predicting h1n1 and seasonal flu: Vaccine cases using ensemble learning approach. pp. 172–

176. In 2020 2nd International Conference on Advances in Computing, Communication Control and Networking (ICACCCN).

Bai, Lu, Jianzhou Wang, Xuejiao Ma and Haiyan Lu. 2018. Air pollution forecasts: An overview. International Journal of Environmental Research and Public Health 15(4): 780.

Bekkar, Abdellatif, Badr Hssina, Samira Douzi and Khadija Douzi. 2021. Air-pollution prediction in smart city, deep learning approach. Journal of Big Data 8(1): 1–21.

Cetin, Mehmet. 2016. A change in the amount of CO_2 at the center of the examination halls: case study of Turkey. Studies on Ethno-Medicine 10(2): 146–155.

Cetin, Mehmet, Ayse Kalayci Onac, Hakan Sevik and Busra Sen. 2019. Temporal and regional change of some air pollution parameters in Bursa. Air Quality, Atmosphere & Health 12: 311–316.

Chattopadhyay, Goutami, Surajit Chattopadhyay and Subrata Kumar Midya. 2021. Fuzzy binary relation-based elucidation of air quality over a highly polluted urban region of India. Earth Science Informatics 14(3): 1625–1631.

Chaurasia, K., Neeraj, B., Burle, D. and Mishra, V. K. 2020a, September. Topographical feature extraction using machine learning techniques from sentinel-2A imagery. pp. 1659–1662. In IGARSS 2020–2020 IEEE International Geoscience and Remote Sensing Symposium.

Chaurasia, K., Tarun, U., Sarala, G. V. and Soni, K. 2020b, November. AI based prediction of daily rainfall from satellite observation for disaster management. pp. 176–188. In SPIE Future Sensing Technologies (Vol. 11525).

Chen, Xiaobing, Lirong Yin, Yulin Fan, Lihong Song, Tingting Ji, Yan Liu et al. 2020. Temporal evolution characteristics of PM2.5 concentration based on continuous wavelet transform. Science of The Total Environment 699: 134244.

Cheng, Yong, Hong Zhang, Zhenhai Liu, Longfei Chen and Ping Wang. 2019. Hybrid algorithm for short-term forecasting of PM2.5 in China. Atmospheric Environment 200: 264–279.

Chowdhury, Mashrur and Adel W. Sadek. 2012. Advantages and limitations of artificial intelligence. Artificial Intelligence Applications to Critical Transportation Issues 6(3): 360–375.

Elangasinghe, Madhavi Anushka, Naresh Singhal, Kim N. Dirks and Jennifer A. Salmond. 2014. Development of an ANN–based air pollution forecasting system with explicit knowledge through sensitivity analysis. Atmospheric Pollution Research 5(4): 696–708.

Elsunousi, Ahlam Ahmed Mohamed, Hakan Sevik, Mehmet Cetin, Halil Baris Ozel and Handan Ucun Ozel. 2021. Periodical and regional change of particulate matter and CO_2 concentration in Misurata. Environmental Monitoring and Assessment 193: 1–15.

Feng, Xiao, Qi Li, Yajie Zhu, Junxiong Hou, Lingyan Jin and Jingjie Wang. 2015. Artificial neural networks forecasting of PM2.5 pollution using air mass trajectory based geographic model and wavelet transformation. Atmospheric Environment 107: 118–128.

Fernando, H. JS楼, Mammarella, M. C., Grandoni, G., Fedele, P., Di Marco, R., Dimitrova, R. et al. 2012. Forecasting PM10 in metropolitan areas: Efficacy of neural networks. Environmental Pollution 163: 62–67.

Ghorani-Azam, Adel, Bamdad Riahi-Zanjani and Mahdi Balali-Mood. 2016. Effects of air pollution on human health and practical measures for prevention in Iran. Journal of Research in Medical Sciences: The Official Journal of Isfahan University of Medical Sciences 21.

Gilik, Aysenur, Arif Selcuk Ogrenci and Atilla Ozmen. 2022. Air quality prediction using CNN+ LSTM-based hybrid deep learning architecture. Environmental Science and Pollution Research: 1–19.

Guo, Qingchun, Zhenfang He, Shanshan Li, Xinzhou Li, Jingjing Meng, Zhanfang Hou et al. 2020. Air pollution forecasting using artificial and wavelet neural networks with meteorological conditions. Aerosol and Air Quality Research 20(6): 1429–1439.

Han, Xiao, Yuqin Liu, Hong Gao, Jianmin Ma, Xiaoxuan Mao, Yuting Wang et al. 2017. Forecasting PM2.5 induced male lung cancer morbidity in China using satellite retrieved PM2.5 and spatial analysis. Science of the Total Environment 607: 1009–1017.

Hashimoto, Daniel A., Elan Witkowski, Lei Gao, Ozanan Meireles and Guy Rosman. 2020. Artificial intelligence in anesthesiology: current techniques, clinical applications, and limitations. Anesthesiology 132(2): 379–394.

Hobday, Alistair J., Lisa V. Alexander, Sarah E. Perkins, Dan A. Smale, Sandra C. Straub, Eric C. J. Oliver et al. 2016. A hierarchical approach to defining marine heatwaves. Progress in Oceanography 141: 227–238.

Hu, Xuefei, Jessica H. Belle, Xia Meng, Avani Wildani, Lance A. Waller, Matthew J. Strickland et al. 2017. Estimating PM2.5 concentrations in the conterminous United States using the random forest approach. Environmental Science & Technology 51(12): 6936–6944.

Inampudi, S., Johnson, G., Jhaveri, J., Niranjan, S., Chaurasia, K. and Dixit, M. 2021. Machine learning based prediction of H1N1 and seasonal flu vaccination. pp. 139–150. In Advanced Computing: 10th International Conference, IACC 2020, Panaji, Goa, India, December 5–6, 2020, Revised Selected Papers, Part I 10. Springer Singapore.

Kaur, Jatinder, Sarbjit Singh and Kulwinder Singh Parmar. 2023. Forecasting of AQI (PM2. 5) for the three most polluted cities in India during COVID-19 by hybrid Daubechies discrete wavelet decomposition and autoregressive (Db-DWD-ARIMA) model. Environmental Science and Pollution Research: 1–18.

Kumari, Rashmi, Subhranil Das and Raghwendra Kishore Singh. 2023. Agglomeration of deep learning networks for classifying binary and multiclass classifications using 3D MRI images for early diagnosis of Alzheimer's disease: a feature-node approach. International Journal of System Assurance Engineering and Management: 1–19.

Liu, Bingchun, Lei Zhang, Qingshan Wang and Jiali Chen. 2021. A novel method for regional NO_2 concentration prediction using discrete wavelet transform and an LSTM network. Computational Intelligence and Neuroscience 2021: 1–14.

Liu, H., Jin, K. and Duan, Z. 2019. Air PM2.5 concentration multi-step forecasting using a new hybrid modeling method: comparing cases for four cities in China. Atmos. Pollut. Res. 10: 1588–1600.

Liu, Hui, Guangxi Yan, Zhu Duan and Chao Chen. 2021. Intelligent modeling strategies for forecasting air quality time series: A review. Applied Soft Computing 102: 106957.

Mannucci, Pier Mannuccio and Massimo Franchini. 2017. Health effects of ambient air pollution in developing countries. International Journal of Environmental Research and Public Health 14(9): 1048.

Masood, Adil and Kafeel Ahmad. 2021. A review on emerging artificial intelligence (AI) techniques for air pollution forecasting: Fundamentals, application and performance. Journal of Cleaner Production 322: 129072.

Mirzadeh, S. M., Nejadkoorki, F., Mirhoseini, S. A. and Moosavi, V. 2022. Developing a wavelet-AI hybrid model for short-and long-term predictions of the pollutant concentration of particulate matter 10. International Journal of Environmental Science and Technology: 1–14.

Mishra, Dhirendra and Pramila Goyal. 2015. Development of artificial intelligence based NO_2 forecasting models at Taj Mahal, Agra. Atmospheric Pollution Research 6(1): 99–106.

Murillo-Escobar, J., Sepulveda-Suescun, J. P., Correa, M. A. and Orrego-Metaute, D. 2019. Forecasting concentrations of air pollutants using support vector regression improved with particle swarm optimization: Case study in Aburrá Valley, Colombia. Urban Climate 29: 100473.

Najjar, Yousef S. H. 2011. Gaseous pollutants formation and their harmful effects on health and environment. Innovative Energy Policies 1(1).

Nilashi, Mehrbakhsh, Hossein Ahmadi, Azizah Abdul Manaf, Tarik A. Rashid, Sarminah Samad, Leila Shahmoradi et al. 2020. Coronary heart disease diagnosis through self-organizing map and fuzzy support vector machine with incremental updates. International Journal of Fuzzy Systems 22: 1376–1388.

Norhayati, I. and Rashid, M. 2018. Adaptive neuro-fuzzy prediction of carbon monoxide emission from a clinical waste incineration plant. Neural Computing and Applications 30: 3049–3061.

Ong, Bun Theang, Komei Sugiura and Koji Zettsu. 2016. Dynamically pre-trained deep recurrent neural networks using environmental monitoring data for predicting PM 2.5. Neural Computing and Applications 27: 1553–1566.

Pardo, Esteban and Norberto Malpica. 2017. Air quality forecasting in Madrid using long short-term memory networks. pp. 232–239. In International Work-Conference on the Interplay Between Natural and Artificial Computation. Cham: Springer International Publishing.

Qader, Mohammed Redha, Shahnawaz Khan, Mustafa Kamal, Muhammad Usman and Mohammad Haseeb. 2021. Forecasting carbon emissions due to electricity power generation in Bahrain. Environmental Science and Pollution Research: 1–12.

Shams, Seyedeh Reyhaneh, Ali Jahani, Saba Kalantary, Mazaher Moeinaddini and Nematollah Khorasani. 2021. The evaluation on artificial neural networks (ANN) and multiple linear regressions (MLR) models for predicting SO_2 concentration. Urban Climate 37: 100837.

Shaziayani, Wan Nur, Ahmad Zia Ul-Saufie, Sofianita Mutalib, Norazian Mohamad Noor and Nazatul Syadia Zainordin. 2022. Classification prediction of PM10 concentration using a tree-based machine learning approach. Atmosphere 13(4): 538.

Shivakumar, Sheethal, K. Aditya Shastry, Simranjith Singh, Salman Pasha, Vinay, B. C. and Sushma, V. 2022. Machine learning-based air pollution prediction. pp. 17–27. In Recent Advances in Artificial Intelligence and Data Engineering: Select Proceedings of AIDE 2020. Springer Singapore.

Slini, Theodora, Kostas Karatzas and Nicolas Moussiopoulos. 2003. Correlation of air pollution and meteorological data using neural networks. International Journal of Environment and Pollution 20(1-6): 218–229.

Sohn, Sang Hyun, Sea Cheon Oh and Yeong-Koo Yeo. 1999. Prediction of air pollutants by using an artificial neural network. Korean Journal of Chemical Engineering 16: 382–387.

Song, Xijuan, Jijiang Huang and Dawei Song. 2019. Air quality prediction based on LSTM-Kalman model. pp. 695–699. In 2019 IEEE 8th Joint International Information Technology and Artificial Intelligence Conference (ITAIC). IEEE.

Subhranil Das and Rashmi Kumari. 2021. Statistical analysis of COVID cases in India. In Journal of Physics: Conference Series, 1797(1): 012006. IOP Publishing.

Subhranil Das, Rashmi Kumari and Deepak Kumar, S. 2021. A review on applications of simultaneous localization and mapping method in autonomous vehicles. Advances in Interdisciplinary Engineering: Select Proceedings of FLAME 2020: 367–375.

Sun, Wei and Jingyi Sun. 2017. Daily PM2. 5 concentration prediction based on principal component analysis and LSSVM optimized by cuckoo search algorithm. Journal of Environmental Management 188: 144–152.

Wang, Chao, Zhirui Ye, Yongbo Yu and Wei Gong. 2018. Estimation of bus emission models for different fuel types of buses under real conditions. Science of the Total Environment 640: 965–972.

Wang, Deyun, Shuai Wei, Hongyuan Luo, Chenqiang Yue and Olivier Grunder. 2017. A novel hybrid model for air quality index forecasting based on two-phase decomposition technique and modified extreme learning machine. Science of the Total Environment 580: 719–733.

Wang, Ping, Yong Liu, Zuodong Qin and Guisheng Zhang. 2015. A novel hybrid forecasting model for PM10 and SO_2 daily concentrations. Science of the Total Environment 505: 1202–1212.

Wang, Yuanni and Tao Kong. 2019. Air quality predictive modeling based on an improved decision tree in a weather-smart grid. IEEE Access 7: 172892–172901.

Wen, Congcong, Shufu Liu, Xiaojing Yao, Ling Peng, Xiang Li, Yuan Hu et al. 2019. A novel spatiotemporal convolutional long short-term neural network for air pollution prediction. Science of the Total Environment 654: 1091–1099.

World Health Organization. 2005. The World Health Report 2005: Make every mother and child count. World Health Organization.

World Health Organization. 2018. WHO report on surveillance of antibiotic consumption: 2016–2018 early implementation.

Xu, Yunzhen, Pei Du and Jianzhou Wang. 2017. Research and application of a hybrid model based on dynamic fuzzy synthetic evaluation for establishing air quality forecasting and early warning system: A case study in China. Environmental Pollution 223: 435–448.

Yan, Dan, Ying Kong, Bin Ye and Haitao Xiang. 2021. Spatio-temporal variation and daily prediction of PM 2.5 concentration in world-class urban agglomerations of China. Environmental Geochemistry and Health 43: 301–316.

Zeng, Yongkang, Jingjing Chen, Ning Jin, Xiaoping Jin and Yang Du. 2022. Air quality forecasting with hybrid LSTM and extended stationary wavelet transform. Building and Environment 213: 108822.

Zhou, Ji-hong, Jun-guang Zhao and Ping Li. 2010. Study on gray numerical model of air pollution in wuan city. pp. 321–323. In 2010 International Conference on Challenges in Environmental Science and Computer Engineering, vol. 1. IEEE.

Zhou, Yanlai, Fi-John Chang, Li-Chiu Chang, I-Feng Kao, Yi-Shin Wang and Che-Chia Kang. 2019. Multi-output support vector machine for regional multi-step-ahead PM2.5 forecasting. Science of the Total Environment 651: 230–240.

Zhu, Furong, Rui Ding, Ruoqian Lei, Han Cheng, Jie Liu, Chaowei Shen, Chao Zhang et al. 2019. The short-term effects of air pollution on respiratory diseases and lung cancer mortality in Hefei: A time-series analysis. Respiratory Medicine 146: 57–65.

3

Groundwater Quality and Human Health Risk Assessment in India

Leveraging GeoAI for Environmental Monitoring

Ravi Prakash Singh

Introduction

Groundwater serves as crucial global water supply (Alley et al. 2002). However, the swift rise in population, industrialization, and unscientific agricultural practices is causing a threat to the quality and quantity of groundwater. As humans rely on freshwater for drinking, industrial use, and agriculture, it becomes imperative to judiciously manage this resource for sustainable development (Mukherjee 2009). Groundwater exists in the soil particles, rocks, and within cracks in bedrocks (Mukherjee 2008). Groundwater quality is influenced by chemical, physical, and biological factors. Due to diverse geological formations, lithological variations and climatic conditions groundwater exhibits a complex occurrence in India. The quality of groundwater depends upon changes in the chemical properties of meteoric water, soil-water interactions, mineral dissolution, duration of

Department of Environmental Sciences, SRM University Delhi-NCR – India; Currently at Department of Environment Science & Technology, Central University of Punjab, Bathinda, Punjab-India. Email: singhravi004@gmail.com soil-water interaction, and human influence (Stallard and Edmond 1983, Subba Rao 2006).

The geochemical composition of groundwater quality is determined by factors such as precipitation composition, mineralogy, topography, and the climate of the study area. These factors collectively give rise to a diverse array of water types that vary spatially and temporally. The distribution of water quality parameters is dependent on hydrochemical processes within the study area (McArthur et al. 2004). Contaminants in groundwater can pose health risks to humans through several processes (He and Wu 2019). The pollution of water by heavy metals is a significant environmental challenge due to its persistence, toxicity, and bioaccumulative properties (Pekey et al. 2004). Consumption of hard water can create the problem of kidney related diseases. Exposure to heavy metals has been associated with serious human health consequences, including skeletal diseases, cardiovascular, infertility, and neural toxicity (WHO 2011). These metals can also cause various liver and kidney disorders and are considered genotoxic carcinogens. Although some metals are essential for the growth and function of living organisms, but higher concentrations of metals can be harmful to human health (Figure 4).

Arsenic Pollution in Groundwater and its Human Health Implications

Arsenic pollution in drinking water is an environmental concern which has affected the health and livelihood of millions of people. In India arsenic contamination in ground water has been reported in the Ganga and Brahmaputra River basin (Das et al. 1996) and several studies have been conducted in West Bengal and Bangladesh to assess the arsenic mobilization in ground water and its correlation with other parameters (McArthur et al. 2004). Arsenic pollution in groundwater is reported in many parts of India (Figure 1) and the source of arsenic in water is mainly geogenic in nature, coming through the reductive dissolution of iron hydroxide (FeOOH), which released the absorbed arsenic in the aquatic environment (McArthur et al. 2004).

The toxicity of arsenic depends on its chemical form. It can be found in four oxidation states such as arsenite (AsIII), arsine (As−III) arsenate (AsV), and arsenic (As) (Sharma and Sohn 2009). In aquatic environments, inorganic arsenic, which includes AsIII, & AsV predominates, while elemental arsenic is uncommon. As−III is mainly present in extremely reduced environmental condition having a low redox potential.

Organometallic compounds exhibit higher toxicity compared to inorganic metals. Elevated levels of arsenic, surpassing the World Health Organization's (WHO) recommended value of 10 µg/L, have been observed in food and drinking water across various regions worldwide (Figure 3). The

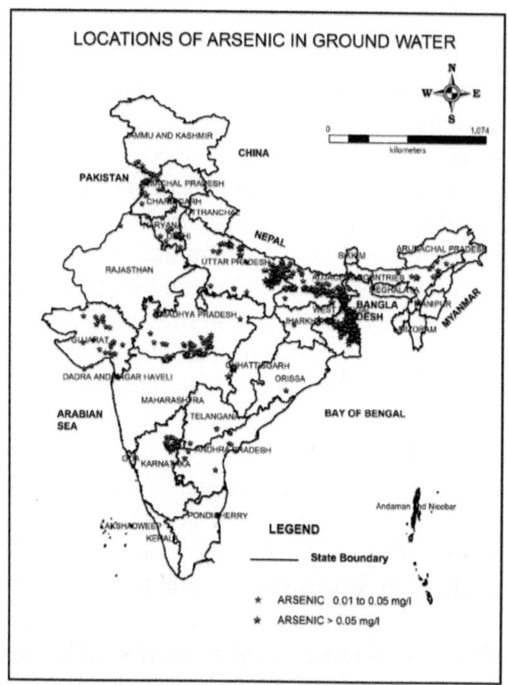

Figure 1. Hotspot of Arsenic in groundwater (CGWB 2014).

chronic exposure to arsenic poses a significant global public health concern. Human exposure to arsenic can originate from both natural sources and human activities through contaminated food and drinking water. Geologic formations, including arsenic accumulation in sedimentary rocks, and soils, represent crucial natural sources. Groundwater contamination is very severe in the aquifers of West Bengal (India), and Bangladesh where millions of populations consume water with arsenic levels exceeding 10 μg/L (Rahman et al. 2021).

Human Health Risk Assessment using Geographic Information System (GIS)

The Geographic Information system uses 3 steps to assess human health risk and acts as an important tool to assess the health risk of humans (Figure 2).

Hazard identification

The hazard level was indicated by the concentrations of arsenic present in drinking water. Anticipated arsenic levels in water were determined through

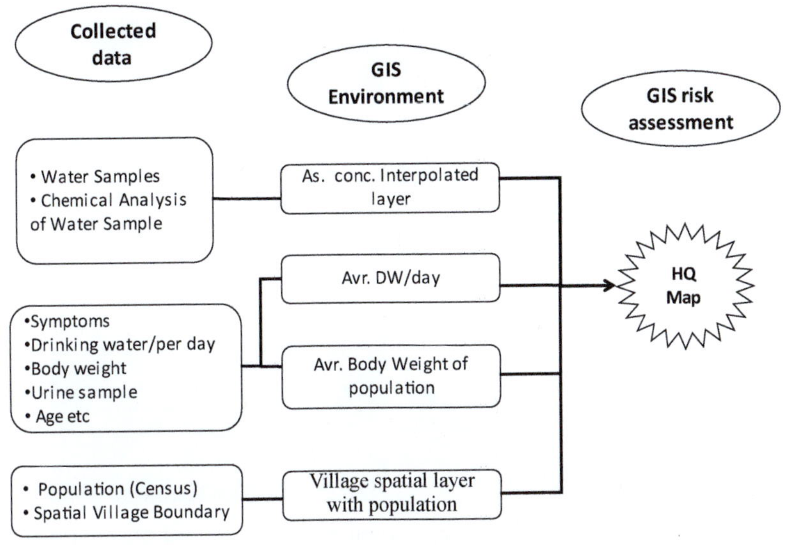

Figure 2. Generation of Hazard Quotient map on GIS platform.

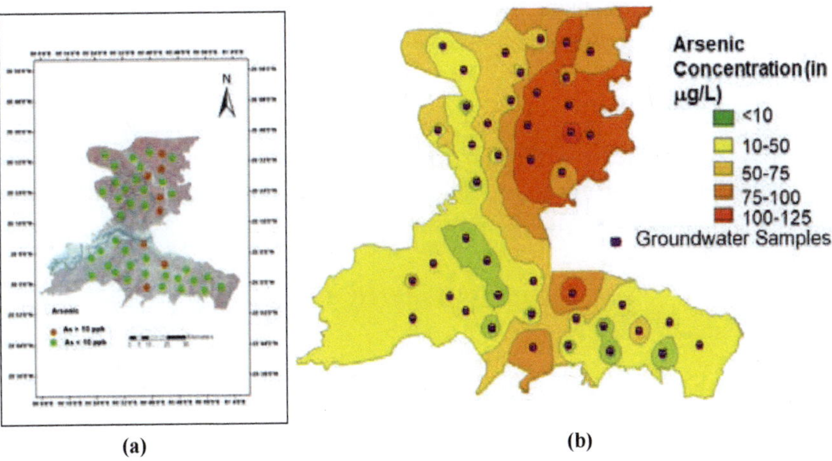

(a)   (b)

Figure 3. (a) Distribution of Arsenic in Bongaigaoun and Goalpara district of Assam. (b) Interpolated Arsenic concentration layer.

inverse distance weighted (IDW) interpolation, utilizing the geographic locations of the observed arsenic concentration in water.

Exposure assessment

Exposure levels were reflected by the concentrations of urinary arsenic (UAs). The anticipated UAs levels for the specified geographic area were acquired

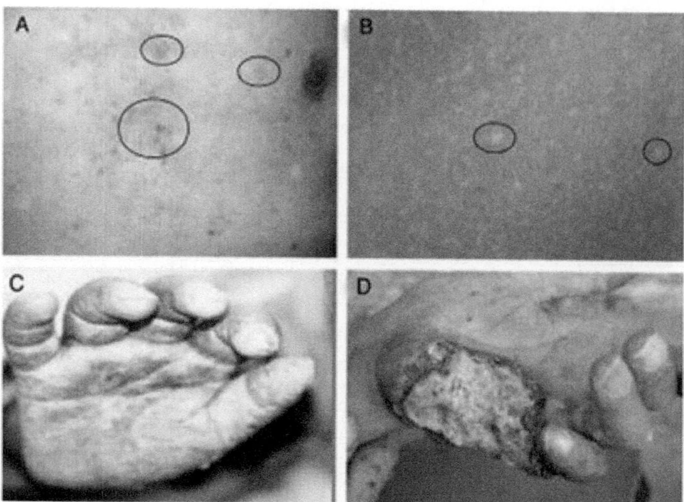

Figure 4. (A) hyperpigmentation, (B) hypopigmentation, (C) keratosis, and (D) skin cancer (Jack et al. 2003).

through interpolation, using the expected levels of arsenic (As) in water along the urine/water correlation curve.

Risk characterization

The risk was indicated by both the hazard quotient and the cancer risk (as shown in equation (1) and equation (2)). calculations for arsenic (As) were based on the anticipated levels of arsenic in water.

$$Hazard\ Quotient\ (HQ) = \frac{DI}{RfD} \tag{1}$$

$$DI = \frac{CW + IR}{BW} \tag{2}$$

DI = Daily intake of arsenic
RfD = Reference dose (0.0003 mg As/KgBw/day)
Cw = Arsenic Concentration in Water (mg/L)
IR = Intake of water (L/day)
Bw = Body Weight (Kg)   HQ > 1 = Health risk situation
                          HQ < 1 = No effect

Fluoride Pollution in Groundwater and its Human Health Implications

Fluoride contamination poses a significant environmental issue in India, with elevated levels widely observed across various regions, notably in Andhra Pradesh, Tamil Nadu, Uttar Pradesh, Gujarat, and Rajasthan. In these states, 50–100% of districts experience excessive fluoride in drinking water sources, primarily stemming from rock-water interaction. The second major concern is arsenic contamination due to its carcinogenic properties, posing risks to residents in affected areas. Fluoride is naturally present in minerals like fluorite, apatite, cryolite, and topaz. Additionally, certain minerals such as biotite, muscovite, and hornblende can contain significant percentages of fluoride. Geogenic fluoride contamination in groundwater is prevalent in India, not only in rocks but also in plants, soil, phosphatic fertilizers, and rock minerals.

The occurrence and hydrogeochemistry of fluoride involve common minerals like fluorite and apatite with low solubility, found in sedimentary and igneous rocks. Amphiboles like hornblende and some micas may also contain fluoride. Although fluoride could associate with silica in acid solutions, such stability is rarely achieved in natural water. Various solubility controls limit the concentration of dissolved fluoride in water. Geogenic fluoride enrichment in groundwater results from leaching and weathering of fluoride-bearing minerals in rocks and sediments, influenced by factors such as water origin, water-bearing medium composition, duration of contact, temperature, pressure conditions, ion-exchange, recharge-discharge rates, and more. From a health perspective, low-level fluoride is beneficial for preventing dental caries, but excessive consumption can lead to severe health issues. Prolonged exposure to groundwater with fluoride levels exceeding 1.5 mg/l results in fluorosis, manifesting as dental, skeletal, and non-skeletal types. Fluoride poisoning causes intense pain, rigidity, and restricted movements in the cervical and lumbar spine, knee and pelvic joints, as well as shoulder joints.

Geospatial Artificial Intelligence

Geospatial Artificial Intelligence (GeoAI) encompasses the fusion of artificial intelligence (AI) methodologies with geospatial data for the examination, comprehension, and extraction of meaningful insights from spatial information. This interdisciplinary realm merges geospatial technologies, including Geographic Information Systems (GIS), with advanced AI approaches to tackle intricate spatial challenges and optimize decision-making processes.

Key components of Geospatial AI

Geospatial Data: GeoAI relies on diverse geospatial data forms, such as satellite imagery, aerial photographs, GPS data, and other location-based information. These data sources furnish a spatial context for AI algorithms to scrutinize.

Machine Learning and Deep Learning: AI methodologies, particularly machine learning (ML) and deep learning, assume pivotal roles in GeoAI. These algorithms can undergo training to discern patterns, categorize land cover (Kuldeep and Garg 2014), recognize objects (Chaurasia et al. 2021a), and forecast changes over time (Chaurasia et al. 2020b) by utilizing geospatial data.

Spatial Analysis: GeoAI streamlines spatial analysis by employing AI algorithms to extract meaningful patterns and relationships from geospatial datasets. This encompasses tasks like spatial clustering, anomaly detection, and recognition of spatial patterns.

Remote Sensing: Remote sensing technologies, such as satellite and aerial imagery, constitute integral components of GeoAI applications. AI algorithms can scrutinize and interpret these images to monitor alterations in the environment, evaluate crop health, trace urban development (Dixit et al. 2021), and more.

Geospatial Modelling: AI models can be harnessed to formulate sophisticated geospatial models for the prediction and simulation of real-world phenomena. These models prove beneficial in areas such as urban planning, environmental impact assessment, disaster response, and other scenarios dependent on spatial considerations.

GeoAI is an emerging field that integrates advancements in spatial data science, artificial intelligence (AI), machine learning (ML), and extensive geospatial datasets (VoPham 2018). It involves the exploration, creation, and implementation of intelligent computer programs designed for the automated processing of both geospatial and non-spatial data. GeoAI encompasses tasks such as modeling geospatial relationships, predicting spatial dynamic phenomena, offering spatial reasoning, and uncovering spatio-temporal patterns and trends (Janowicz 2020).

The methods employed in GeoAI leverage AI and ML techniques for geospatial modeling, including applications such as fluvial landform classifications and spatial hydrological prediction. These GeoAI and ML methods fall into categories such as unsupervised learning supervised learning and the optimization of modelling problems.

Application of Geospatial Artificial Intelligence in Groundwater Quality

Traditionally, the classical conceptual and numerical groundwater models have been employed to forecast the dynamic changes in hydrogeological conditions and processes, drawing upon a deep understanding of observed behaviors. The accuracy and reliability of model predictions depend mainly on sufficient and appropriate data for the calibration and validation of the hydrological system (Gao and Li 2014). Although these methods face limitations, like the requirement of a large number of input parameters throughout the modeling process (Figure 5). Moreover, many models have limitation to effectively handle and quantify predictive hydrogeological non-linearity and uncertainties. Limitation to recognize and evaluate these non-linearities and uncertainties can produce an inaccurate representation of the actual system, contributing to poor groundwater model performance and a decline in accuracy in forecasting (Guzman et al. 2015). This chapter extensively explores the new techniques and implications of the most powerful artificial intelligence (AI) approaches in a succinct and integrated manner, particularly focusing on their application in modeling and forecasting groundwater quality for domestic use (Figure 6). The study delves into four widely utilized AI methods such as artificial neural network (ANN), adaptive network-based fuzzy inference system (ANFIS), evolutionary algorithm (EA), and support vector machine (SVM). It scrutinizes their features and capabilities while identifying the key challenges inherent in achieving desired results. Upon analysis of all four AI methods, it is observed that ANN outperforms others when handling large

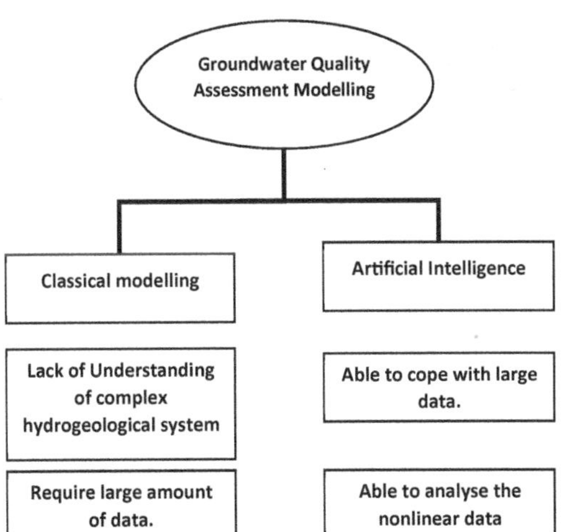

Figure 5. Classical Vs AI based groundwater quality modelling.

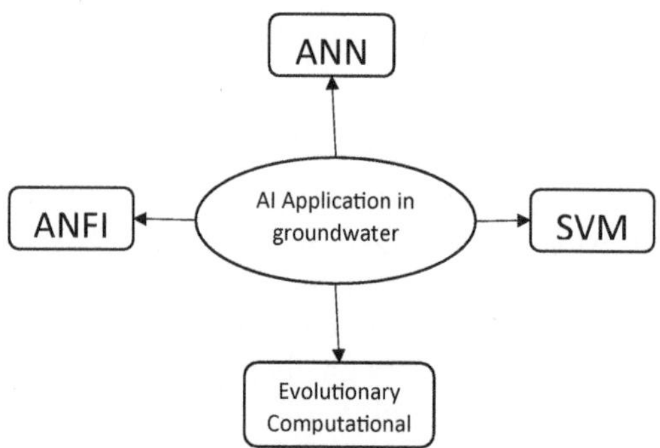

Figure 6. Artificial Intelligence (AI) techniques in groundwater quality assessment.

datasets, demonstrating precise predictions by virtue of its capacity to model intricate non-linear and complex relationships, despite certain limitations.

The review findings underscore that the effective adoption of AI models is influenced by considerations such as the appropriateness of input parameters, the nature of individual functions, and the efficiency of performance metrics.

The insights gained from this study are valuable for the development of the groundwater plans and contribute to the enhancement of AI applications in ensuring groundwater quality and quantity. The study concludes by presenting recommendations aimed at advancing knowledge development to improve modeling structures in the specified domain.

Artificial neural network

Artificial neural network (ANN), serves as a simplified representation of a mathematical model inspired by the structure of a biological neural network, mimicking human cognitive abilities (Fausett 1994) distinguishes neural networks based on factors such as connection patterns, node architecture, methods of configuring connection weights, and the activation function. In the configuration of ANN models, there are at least 3 layers of interconnected neurons, encompassing one or more hidden layers, an output layer, and an input layer as depicted in Figure 7 (Nur et al. 2021). The parallel distributed processor of ANNs processes information from input to output through a network structure comprising interconnected nodes. The input data undergo computation in the input layer, and the output layer corresponds to the network's response based on existing databases (Kheradpisheh et al. 2015). The intermediate hidden layer assumes a crucial role in representing and computing complex associations between patterns.

Among the diverse types of artificial neural networks (ANNs), the back-propagation neural network (BPNN), Levenberg-Marquardt back-propagation (LMBP), and multilayer perceptron (MLP) algorithms are

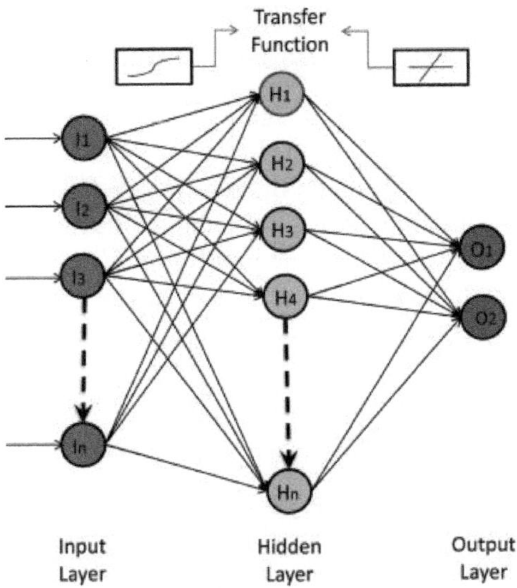

Figure 7. A typical ANN with input, output and hidden layer (Nur et al. 2021).

widely used in groundwater modeling due to their accuracy. The feed-forward neural network (FFNN) is a straightforward algorithm where information flows unidirectionally from input nodes through hidden nodes to output nodes.

Researchers actively utilize ANNs in water resources management for various applications, including water and groundwater level and quality forecasting, sediment modeling, and rainfall-runoff estimation (Shiri et al. 2013). The ANN is widely applied in groundwater modelling due to its effectiveness in predictive hydrochemical and hydrogeological parameters (Yu and Liu 2012).

In the existing study the application of hidden neuron numbers and learning performance for different input variables concerning saltwater intrusion and arsenic pollution is also evaluated. Input variables for saltwater intrusion included electrical conductivity, Cl^-, SO_4^{2-}, K^+, and Mg^{2+}, while variables for arsenic pollution were alkalinity arsenic (As), and total organic carbon. It was also observed that the performance of ANN was unaffected by a higher number of hidden neurons, but finding the optimal number of hidden layers is very crucial for generalization (Mezard and Nadal 1989). Accurate and reliable forecasts rely on both current and previous input data.

Various authors have employed the random forest method to forecast the distribution of arsenic concentration in groundwater. Utilizing a dataset comprising arsenic concentration and predictor variables, a statistical prediction model was developed to anticipate instances where arsenic levels in groundwater surpass the WHO and national India guideline concentration of 10 µg/L for drinking water. Subsequently, a hazard risk map was generated to facilitate effective management of safe drinking water.

Conclusion

The increased levels of arsenic and fluoride in groundwater pose detrimental effects on human health. Integrating artificial intelligence (AI) methodologies with geospatial data offers a valuable tool for extracting meaningful insights from spatial information. This fusion can serve as a crucial means to predict groundwater quality and generate probability hazard maps, aiding in effective water resource management.

Acknowledgement

The author is thankful to SRM University Delhi-NCR for supporting and enabling the research project.

References

Alley, W. M., Healy, R. W., LaBaugh, J. W. and Reilly, T. E. 2002. Flow and storage in groundwater systems. Science 296: 1985–1990.

Central Groundwater Board Report. 2014. Concept note on geogenic contamination of ground water in India.

Chaurasia, K., Nandy, R., Pawar, O., Singh, R. R. and Ahire, M. 2021a. Semantic segmentation of high-resolution satellite images using deep learning. Earth Science Informatics 14(4): 2161–2170.

Chaurasia, K., Tarun, U., Sarala, G. V. and Soni, K. 2020b, November. AI based prediction of daily rainfall from satellite observation for disaster management. pp. 176–188. In SPIE Future Sensing Technologies (Vol. 11525).

Das, D., Samanta, G., Mandal, B. K., Chowdhury, T. R., Chanda, C. R., Chowdhury, P. P. et al. 1996. Arsenic in groundwater in six districts of West Bengal, India. Environmental Geochemistry and Health 18(1): 5–15.

Dixit, M., Chaurasia, K. and Mishra, V. K. 2021. Dilated-ResUnet: A novel deep learning architecture for building extraction from medium resolution multi-spectral satellite imagery. Expert Systems with Applications 184: 115530.

Fausett, L. V. 1994. Fundamentals of Neural Networks: Architectures, Algorithms, and Applications Prentice-Hall, Upper Saddle River, NJ, USA.

Gao, L. and Li, D. 2014. A review of hydrological/water-quality models. Front. Agric. Sci. Eng. 267.

Guzman, J., Shirmohammadi, A., Sadeghi, A., Wang, X., Chu, M. L., Jha, M. et al. 2015. Uncertainty considerations in calibration and validation of hydrologic and water quality models. Trans. ASABE (Am. Soc. Agric. Biol. Eng.) 1745–1762.

He, X., Wu, J. and He, S. 2019. Hydrochemical characteristics and quality evaluation of groundwater in terms of health risks in Luohe aquifer in Wuqi County of the Chinese Loess Plateau, northwest China. Hum Ecol. Risk Assess 25: 32–51.

Jack C. Ng, Jianping Wang and Amjad Shraim. 2003. A global health problem caused by arsenic from natural sources. Chemosphere 52(9): 1353–9.f.

Janowicz, K., Gao, S., McKenzie, G., Hu, Y. and Bhaduri, B. 2020. GeoAI: Spatially explicit artificial intelligence techniques for geographic knowledge discovery and beyond. Int. J. Geogr. Inf. Sci.

Kheradpisheh, Z., Talebi, A., Rafati, L., Ghaneian, M. T. and Ehrampoush, M. H. 2015. Groundwater quality assessment using artificial neural network: a case study of Bahabad plain, Yazd, Iran. Desert 20(1): 65–71.

Kuldeep and Garg, P. K. 2014, Texture based information extraction from high resolution images using object-based classification approach. pp. 299–303. In 2014 Third International Workshop on Earth Observation and Remote Sensing Applications (EORSA).

McArthur M., Banerjee, D. M, Hudson-Edwards, K. A., Mishra, R., Purohit, R., Ravenscroft, P. et al. 2004. Natural organic matter in sedimentary basins and its relation to arsenic in anoxic ground water: the example of West Bengal and its worldwide implications. Applied Geochemistry 19: 1255–1293.

Mezard, M. and Nadal, J. P. 1989. Learning in feedforward layered networks: the tiling algorithm. J. Phys. A: Math. Gen. 22(12): 2191–2203.

Md. Mostafizur Rahman, Nathan Mise, Md. Tajuddin Sikder, Gaku Ichihara, Md. Khabir Uddin, Masaaki Kurasaki et al. 2021. Environmental arsenic exposure and its contribution to human diseases, toxicity mechanism and management. Environmental Pollution 289.

Mukherjee, S. 2008. Role of Satellite sensors in groundwater exploaration. Sensors 8(3): 2006–2016.

Mukherjee, S. 2009. Sensible measures to guard India's groundwater supply. Nature 462: 276.

Nur Farahin Che Nordin, Nuruol Syuhadaa Mohd, Suhana Koting, Zubaidah Ismail, Mohsen Sherif and Ahmed El-Shafie. 2021. Groundwater quality forecasting modelling using artificial intelligence: A review. Groundwater for Sustainable Development 24.

Pekey, H. 2004. The distribution and sources of heavy metals in Izmit Bay surface sediments affected by a polluted stream. Marine Pollution Bulletin 52: 1197–1208.

Sharma Virendra and Sohn Mary. 2009. Aquatic arsenic: toxicity, speciation, transformations, and remediation. Environmental International 35(4): 743–59.

Shiri, J., Kisi, O., Yoon, H., Lee, K.-K. and Hossein Nazemi, A. 2013. Predicting groundwater level fluctuations with meteorological effect implications—a comparative study among soft computing techniques. Computers and Geoscience 56: 32–44.

Stallard, R. F. and Edmond, J. M. 1983. Geochemistry of the Amazon, the influence of geology and weathering environment on the dissolved load. Journal of Geophysical Research 88: 9671–9688.

Subba Rao, N. 2006. Seasonal variation of groundwater quality a part of Guntur District, Andhra Pradesh, India. Environmental Geology 49: 413–429.

VoPham, T., Hart, J. E., Laden, F. and Chiang, Y.-Y. 2018. Emerging trends in geospatial artificial intelligence (GeoAI): Potential applications for environmental epidemiology. Environ. Health 17: 40.

World Health Organization Report. 2011. Arsenic in Drinking Water. WHO/SDE/WSH/03.04/75/ Rev/1.

Yu, F. R. and Liu, Z. P. 2012. The application of artificial neural network in the groundwater quality assessment in industrial Park. Catchment Adv. Mat. Res. 518–523: 1340–1343.

4

An Integrated Stratagem for Soil Pollution Assessment Utilizing Geospatial Tools and Machine Learning Approach

Ghazaala Yasmin,[1,*] *Kuldeep Chaurasia*[2] and *Padam Jee Omar*[3]

Introduction

A healthy environment, economic growth, and increased agricultural output have been severely hampered by the global decline in soil quality. Soil degradation weakens the quality of the soil by causing erosion. The unfavourable changes in salinity, acidity, or alkalinity, debt of biological stuff, soil potency cutback, anatomical disintegration is consequences hazardous actinic (Garg 2015, Jia 2021). The harmful chemical resource reconciliation consequence soil degradation. It can also be accountable for provoking social, economic problems. Taking as illustration like land acquisition, capital, and infrastructure, this consequence an immense downturn in the chemical trait and physical attributes of the soil resources. Soil contamination is an environmental disaster that has an influence on ecosystems, agricultural output, and public wellbeing. Heavy metals, pesticides, industrial chemicals, and other toxins

[1] Department of Computer Science and Engineering/Information Technology, Jaypee Institute of Information Technology, Uttar Pradesh-201307.

[2] School of Computer Science Engineering and Technology, Bennett University, Uttar Pradesh-201310.

[3] Department of Civil Engineering, Babasaheb Bhimrao Ambedkar University, Lucknow-226025.

Emails: kchaurasia.iitr@gmail.com; drpadamomar@gmail.com

* Corresponding author: ghazaala.yasmin@gmail.com

have found refuge in soil due to growing industry, urbanisation, and intense agricultural practises. As far as spatial distribution is concerned, the magnitude, and dynamics of soil pollution is vitally important for effective mitigation strategies and sustainable land management practices. Geographic information systems (GIS) provide unrivalled capabilities in spatial analysis, visualisation, and modelling (Kuldeep et al. 2017). They have become an indispensable tool in environmental research. If we contextualize soil pollution there are a diverse number of sources which pollute soil. They range from industrial activities, inappropriate waste disposal, mining operations, agricultural bolt to atmospheric deposition. These activities cause an assortment of contaminants to enter soil, which has long-term negative consequences on biodiversity, soil fertility, and human health. Developing focused repair techniques requires an understanding of the intricate interactions between these sources and their spatial distribution.

Motivation

The well-being of ecological systems is directly impacted by soil pollution, which effects the existence of animals and plants. Understanding the level of contaminants and its effect on biodiversity is made easier by analysing soil pollution. Food safety can be impacted by soil pollution and agricultural yields can be significantly reduced by its contamination. By identifying contaminated regions and facilitating focused rehabilitation, pollution analysis helps ensure safer food production. Human health can be seriously jeopardised by polluted soil, particularly if the contaminants seep into groundwater or the food cycle. By analysing soil pollution, we gain insights into the causes, effects, and potential solutions. Machine learning helps to take relevant decisions, implement effective strategies for pollution control, and ensure a healthier environment for current and future generations. Applying GIS and GPS technology makes the detection of soil pollution early and quick. This will help in producing better quality crops which will boost the economy of the country and may help in balancing the ecosystem.

Contribution

This proposed chapter seeks to use GIS and GPS for the analysis of soil pollution. The overall system has been implemented by intruding the following major contribution,

- The identification of important pollutants and the spatial study of soil contamination hotspots in specific geographical locations using GPS.
- GIS is an effective predictive model to estimate the distribution and dynamics of soil pollution over time.

- The proposed chapter utilises GPS-enabled devices to record the precise positions where soil samples are taken from various places and area. These may consider factors such as topography, land utilization and proximity to pollution sources.

- To make the system data more robust multidisciplinary datasets are being gathered to justify versatile cases of soil contamination.

- An effective machine learning based decision-making tools has been used to test the efficiency of the system.

- Many other statistical measures have been considered to get a better result.

Literature Survey

Analysis of soil pollution techniques is a vast area for researchers. The existing work encompasses an exhaustive array of studies to investigate the impacts, causes, remedies and maintenance of soil contaminants. The proposed chapter highlights a brief review to summarize various facets of research regarding soil pollution. Various researchers have proposed a method on the work related to soil pollution analysis and prediction. Gao et al. (Gao et al. 2022) have proposed spatial prediction of soil pollution using machine learning approach. Marian et al. (Marian et al. 2022) has proposed an effective way of a spatio-temporal based prediction on soil moisture using Soil Moisture Active Passive (SMAP) retrieval and topographic indices. Similarly, these topographic indices are also helpful for finding remote vegetation area for an index-based gross primary productivity (GPP) estimation based on watershed scale (Xie 2022). In order to classify the land cover, an effective attentive geometric feature pyramid network has been implemented. This technique helps in finding the area where the pollution of soil variation can be analysed (Li et al. 2022). The analysis of soil pollution is also done using drone image recognition for a very huge area or remotely located area on the field such as arsenic-contaminated agricultural field (Jia 2021). The topographic indices can map with global positioning system in order to understand geographic information system (Heiniger 1991, Karr 2011). Heavy metal pollution-based soil has been predicted by applying a machine learning framework by training the data using machine learning approach (Jia 2019). A novel idea has been adopted by past researchers to detect pollution in soil by incorporating infrared reflectance spectroscopy (VIRS) based methods and combining the features training using machine learning approach (Srivastava 2012). Some of the researchers have incorporated the concept of Geographical Information System [GIS] to monitor pesticide pollution and its effect on public health (Kaminska 2004). Soil contamination is also analysed where the contamination has occurred due to cutting stones in countries like Jordan (Al-Joulani 2008). For many problems related to pollution, Geographical Information System as well as multivariate analysis is also applied in order to tackle contaminants such as water pollution in huge areas (Al-Akhras 2015, Al-Joulani Nabil 2008). Global Positioning Systems (GPS) is also used for air monitoring and air pollution estimation by mapping temperature data to create time-activity classifications (Nethery 2014).

Propound Methodology

The proposed idea is to predict soil pollution using two major techniques (a) feature selection techniques and (b) GIS and GPS spatial mapping data. Initially, the input data containing soil samples are pre-processed. These samples contain chemical compositions such as heavy metals and physical factors such as temperature, moisture, agricultural activities including pesticides, herbicides, and fertilizers. Also land such urbanization and heavy construction or mining area considered are considered as pollutant feature. The features based on the mentioned pollutant features have been pre-processed by applying data mining techniques such as normalization and data integration (Jia 2019). The pre-processed data has been executed on a well-known feature selection technique to get the optimal trait set. These optimal set of features reduces the performance time of the system and increases the accuracy of the system. To select features a combined data of input sample and data through spatial mapping has been executed. Finally, the obtained final set of selected features has been trained to get the accuracy of the performance of the proposed system. The overall layout of the proposed system has been inscribed in Figure 1. The prospective chapter is named as Soil Pollution Analysis Through Machine Learning (SPAM).

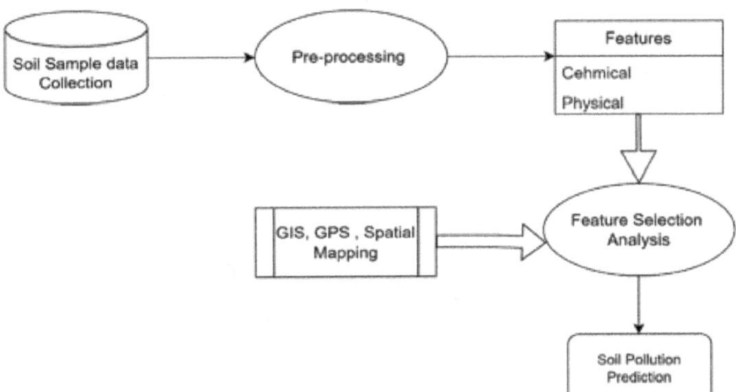

Figure 1. Layout diagram for the propound methodology.

Role of GIS and GPS in Soil Pollution Analysis

GIS is an effective means for collecting, analysing, and visualising data on soil attributes, land use, pollution sources, and pollutant concentrations. GIS enables the construction of contamination maps, identification of hotspots, and the assessment of pollutant shipping over time by integrating soil sampling data with geographical and environmental aspects (Kaminska 2004). Add to that, GIS helps with the identification of sensitive locations as well as the calculation of possible dangers connected with soil pollution, resulting in more informed decisions for land use planning and remediation initiatives.

Moreover, to analyze soil pollution GPS technology is introduced in order to collect exact geographic information related to environmental parameters, land usage, sources of contamination and sample of soil. Use of GPS integration improves soil contamination understanding, mapping, and management by providing precise spatial data.

Spatial Mapping

It is possible to create spatial maps that show the distribution and concentration levels of contaminants throughout the target area by integrating GPS data with the soil sample data (Nethery 2014). In order to visualise pollution areas of greatest need, hues, and possible contamination patterns, GIS software can be used to overlay GPS coordinates of sample points with concentrations of pollutants data. In the proposed method GPS data is integrated with other factors such as precipitation, temperature, type of soil and vegetation cover.. These factors have been incorporated in order to understand the impact of environmental conditions on the behaviour of pollutants.

Experimental Result

Dataset collection

The dataset used in the proposed chapter is composed of two types. One is a sampled dataset collected from different websites which contains metal composition percentage in different agricultural land across India. This dataset is the collection of 1564 sample of data containing different attributes like percentage of metals, moisture level and heavy metals. The most trivial found inorganic contaminants are trace elements such as arsenic, chromium,, copper, mercury, lead, nickel, zinc and many others. Radionuclides pollutants are also part of these contaminant. We have named those as "SamData".

The other dataset is the standard dataset taken for soil pollution from U.S. soil jurisdictions on the basis of the top 100 concerned pollutants (Li et al. 2018). The primary groups of chemicals being identified are elements, such as, cyanides, halogenated methanes, chloroethanes and choroethenes, benzenes, phenols, cancer-causing PAHs, 8 noncarcinogenic PAHs, chemical pesticides that have been used in past times, herbicides that are currently in use, and additional contaminants. We have named that data as "SData".

Another dataset with heavy metal component and their percentage of composition of other country like China have been incorporated in the result analysis (Cao 2021). The experiment has been conducted on farmland soils with heavy metal content in six well known urban areas in Wuhan, Hubei Province, China. The data was obtained from the Wuhan Academy of Agricultural Sciences. As far as the land area and crop layout is concerned in each district, we have considered the type of soil as well as the sampling through GPS locator. The dataset is the total collection of 1161 sets of data. This consists of attributes such as such as longitude, latitude, altitude, functional area and eight different soil heavy metal contents. We have named that data as "CData".

Feature selection

The overall system has been trained on these three datasets. For computing better performance in optimal time of computation, Different well-known feature selection algorithm has been applied those algorithm are Correlation-based Feature Selection (CFS), Consistency based feature selection (CON), some graph based feature selection algorithms, such as Louvain (LV), Infomap (IM), Fastgreedy (FG), Girvan-Newman (GN) (Yasmin 2020), and Rough-spanning tree based feature selection algorithm (RMST) (Yasmin 2022) has been applied to get the most optimal attributes.

The selected feature has been computed on the mentioned three datasets. Then Well-known Classification algorithm has been applied. Eight different classifiers, named as Support vector machine (S), Knearest neighbors (K), Decision tree (D), Neural network (Net), Random forest (RnF), Naïve Bayes (NBa), Adaboost (AdB), and Sequential minimal optimization (Smopt) are used to analyse the accuracy of the system. SVM is computed by setting RBF kernal, KNN value computed through sample size square root in dataset and validation of 10-fold has been applied to measure the performance of the classifiers. The above algorithm has been executed on wekatool platform. The value for the percentage of selected feature corresponding to the feature selection model has been detailed in Table 1.

The performance of the feature selection model has been measured not only in terms of percentage of selected features but also through statistical measure as mentioned. Table 2 is the detailed representation of the numerical figure for all the selected feature selection model mentioned in the proposed work applied of the given datasets.

Table 1. Selected feature by applying well-known attribute exception methodology.

Dataset	Approach for picking traits (elected attribute in %)						
	CFS	CON	GN	IM	LV	FG	RMST
SamData	58	47	36	41	45	47	42
SData	54	51	**31**	39	46	43	40
CData	49	44	32	37	45	42	39

Comparative analysis

The proposed method has been compared with state of art work on soil pollution analysis. Gao (Gao 2022) has used machine learning approach by simplifying artificial intelligence strategy for the analysis of contaminants in soil leveraging spatial forecasting. A very useful technique is being used by Jia, Xiyue (Jia 2021) for Assessing soil contamination at a farming site affected with arsenic employing drone reconnaissance and machine learning. The data used in their work has been used to test our system. The documentary of Jia, Xiaolin (Jia 2019) has given case studies on methodologies framework for the identification of heavy metal contaminant in soil in which Bivariate spatial correlation analysis is one the important analysis. The dataset analysed in their documentary is a versatile data which covers both physical factors in a major role. The ideology of Geographical Information System has been used in the work of Kaminska (Kaminska 2004) which has introduced monitoring of pesticide pollution and impact on public health which occurs due to pollution in salt. The work of Nethery, Elizabeth (Nethery 2014). has been compared to put a significant impact of the proposed work in terms of Using Global Positioning Systems (GPS) data.

The above-mentioned existing work has been compared on the proposed method to classify the performance of the propound idea in the prospective chapter, the comparison has been obtained on three datasets mentioned in the proposed method. The comparison has been summarized in Table 3.

Conclusion

A machine learning driven examining and analysis of soil pollution provides a promising approach to properly understand, anticipate, and alleviate environmental risks. The proposed methodology helps us make better decisions about handling soils and remediation approaches by using algorithms to process large datasets, find trends, and predict contamination levels. Through the utilisation of machine learning models, rehabilitation operations may be optimised, early pollution detection made easier, and sustainable techniques for maintaining soil health can be developed. However, this method can be

Table 2. Performance classification of the proposed method.

Dataset	Feature Selection Algorithm	Classification Algorithm							
		S	K	D	Net	RnF	NBa	AdB	Smopt
SamData	CFS	75.31	76.04	78.29	81.72	83.00	76.20	78.33	77.59
	CON	76.33	77.29	76.50	82.06	82.93	78.37	79.11	76.65
	GN	79.66	80.74	82.30	86.08	89.91	83.16	84.34	80.22
	IM	75.65	74.66	76.10	79.82	81.72	77.29	78.01	78.64
	LV	76.33	74.01	76.08	81.53	79.09	80.87	81.56	79.40
	FG	77.28	75.23	80.79	81.44	80.81	82.01	80.10	80.19
	RMST	79.01	76.73	80.91	79.53	78.55	82.69	81.47	79.38
SData	CFS	78.32	77.95	79.33	81.63	82.45	83.98	83.78	82.22
	CON	80.60	81.51	80.30	82.50	82.91	82.67	83.48	81.77
	GN	81.58	82.62	83.97	83.88	83.01	83.56	82.59	82.69
	IM	79.83	80.51	81.27	82.09	83.40	80.64	84.93	80.11
	LV	79.83	81.01	82.16	81.98	83.58	86.72	83.02	81.06
	FG	78.98	80.50	83.23	80.99	82.65	84.34	81.29	81.55
	RMST	80.62	81.33	82.84	81.49	82.40	83.38	80.63	80.78
CData	CFS	77.99	78.11	78.32	83.22	80.36	81.87	80.61	81.79
	CON	80.23	82.64	81.30	82.00	81.67	83.06	82.22	80.85
	GN	82.66	82.94	82.72	84.94	82.66	82.01	83.03	81.89
	IM	78.40	79.93	80.50	80.13	81.00	83.42	81.91	82.62
	LV	80.38	80.79	81.57	82.38	82.63	82.28	82.08	82.22
	FG	82.52	82.98	84.69	82.56	80.19	81.57	81.92	81.71
	RMST	81.34	83.67	82.47	82.68	81.70	82.04	82.61	81.93

Table 3. Comparison of the suggested approach with cutting-edge work in percentage (%) terms.

Method	SamData	SData	CData
Gao et al. [1]	83.63	84.01	82.59
Jia, Xiyue et al. [5]	82.44	83.07	82.86
Jia, Xiaolin et al. [7]	81.64	81.92	80.30
Kaminska et al. [11]	82.55	83.31	82.87
Nethery, Elizabeth et al. [15]	80.99	81.24	81.10
Proposed SPAM	83.65	83.90	83.05

enhanced by applying other statistical measures in order to check versatile set of soil data. The proposed method can be enhanced by incorporating a hybrid model of topographic mapping and combining this to the GIS and GPS to get more accurate result.

References

Al-Akhras, M. A., Albiss, B. A., Alqudah, M. S. and Odeh, T. S. 2015. Environmental pollution of cell-phone towers: Detection and analysis using geographic information system. Jordan Journal of Earth and Environmental Sciences 7.2: 77–85.

Al-Joulani, Nabil. 2008. Soil contamination in Hebron District due to stone cutting industry. Jordan Journal of Applied Science 10: 37–50.

Cao, Wenqi and Cong Zhang. 2021. Data prediction of soil heavy metal content by deep composite model. Journal of Soils and Sediments 21: 487–498.

Gao, Bingbo, Alfred Stein and Jinfeng Wang. 2022. A two-point machine learning method for the spatial prediction of soil pollution. International Journal of Applied Earth Observation and Geoinformation 108: 102742.

Garg, P. K. 2015. The role of satellite derived data for flood inundation mapping using GIS. The International Archives of the Photogrammetry, Remote Sensing and Spatial Information Sciences 40: 235–239.

Heiniger, Ronnie W. 1999. Understanding geographic information systems and global positioning systems in horticultural applications. HortTechnology 9.4: 539–547.

Jia, X., Hu, B., Marchant, B. P., Zhou, L., Shi, Z. and Zhu, Y. 2019. A methodological framework for identifying potential sources of soil heavy metal pollution based on machine learning: A case study in the Yangtze Delta, China. Environmental Pollution 250: 601–609.

Jia, X., Cao, Y., O'Connor, D., Zhu, J., Tsang, D. C., Zou, B. et al. 2021. Mapping soil pollution by using drone image recognition and machine learning at an arsenic-contaminated agricultural field. Environmental Pollution 270: 116281.

Jia, X., O'Connor, D., Shi, Z. and Hou, D. 2021. VIRS based detection in combination with machine learning for mapping soil pollution. Environmental Pollution 268: 115845.

Kaminska, Iwona A., Anna Oldak and Waldemar A. Turski. 2004. Geographical information system [GIS] as a tool for monitoring and analysing pesticide pollution and its impact on public health. Annals of Agricultural and Environmental Medicine 11.2.

Kuldeep, Banu, V., Uniyal, S. and Nagaraja, R. 2017. Space based inputs for health service development planning in rural areas using GIS. Geodesy and Cartography 43.1: 28–34.

Li, D., Shen, X., Guan, H., Yu, Y., Wang, H., Zhang, G. et al. 2022. AGFP-Net: Attentive geometric feature pyramid network for land cover classification using airborne multispectral LiDAR data. International Journal of Applied Earth Observation and Geoinformation 108: 102723.

Li, Wenbiao, Zijian Li and Aaron Jennings. 2018. Regulatory performance dataset constructed from US soil jurisdictions based on the top 100 concerned pollutants. Data in Brief 21: 36–49.

Nethery, E., Mallach, G., Rainham, D., Goldberg, M. S. and Wheeler, A. J. 2014. Using Global Positioning Systems (GPS) and temperature data to generate time-activity classifications for estimating personal exposure in air monitoring studies: an automated method. Environmental Health 13: 1–11.

Schönauer, M., Prinz, R., Väätäinen, K., Astrup, R., Pszenny, D., Lindeman, H. et al. 2022. Spatio-temporal prediction of soil moisture using soil maps, topographic indices and SMAP retrievals. International Journal of Applied Earth Observation and Geoinformation 108: 102730.

Srivastava, P. K., Han, D., Gupta, M. and Mukherjee, S. 2012. Integrated framework for monitoring groundwater pollution using a geographical information system and multivariate analysis. Hydrological Sciences Journal 57.7: 1453–1472.

Xie, X., Tian, J., Wu, C., Li, A., Jin, H., Bian, J. et al. 2022. Long-term topographic effect on remotely sensed vegetation index-based gross primary productivity (GPP) estimation at the watershed scale. International Journal of Applied Earth Observation and Geoinformation 108: 102755.

Yasmin, G., Das, A. K., Nayak, J., Pelusi, D. and Ding, W. 2020. Graph based feature selection investigating boundary region of rough set for language identification. Expert Systems with Applications 158: 113575.

Yasmin, G., Das, A. K., Nayak, J., Vimal, S. and Dutta, S. 2022. A rough set theory and deep learning-based predictive system for gender recognition using audio speech. Soft Computing 1–24.

5

Statistical Approach to Evaluate Spatio-Temporal Relationship of Crop Residue Burning and Land Surface Temperature Over a Decade

Case Study of Punjab

Amritpal Digra,[1,]* *Satyam Kumar,*[2] *Srimadhi K.,*[1] *Akash Goyal*[3] and *Sameer Saran*[4]

Introduction

The Argo industry plays an important role in the growth of global economy, and Indian economy is dominated by the agricultural sector as half of the Indian population depends on agriculture for livelihood. In the state of Punjab, Haryana and western Uttar Pradesh, a change in farming systems

[1] Research Scientist, Regional Remote Sensing Centre-North, National Remote Sensing Centre, ISRO-110049.

[2] Student, Birla Institute of Technology, Mesra, Jharkhand 835215.

[3] Scientist-SE, Regional Remote Sensing Centre-North, National Remote Sensing Centre, ISRO-110049.

[4] Deputy General Manager & Scientist-SG, Regional Remote Sensing Centre-North, National Remote Sensing Centre, ISRO-110049.

Emails: myselfsatyam19@gmail.com; madhusri3773@gmail.com; akash_g@nrsc.gov.in; sameer_s@nrsc.gov.in

* Corresponding author: amritpaldigra30@gmail.com

occurred from conventional crops to extensive rice-wheat cropping system, resulting in large amounts of agriculture residue. The time gap between these two crops is just 15–20 days. The practice started after 1990s, and government also restricted crop burning when the mechanization of agriculture was taking momentum. The production of residue was highest in Uttar Pradesh, followed by Punjab and Haryana, but when it comes to burning Punjab tops the list followed by Uttar Pradesh (Anuradha et al. 2021).

Wheat straws are generally removed as cattle fodder, but as far as rice/paddy is concerned the straw of these pose a challenge for disposal. Since, the time taken for the disposal is too long, the simple way to get rid of it is by burning it in the field which is termed as Crop Residue Burning (CRB). This happens during the month of April–May and November–December. The major results of crop residue burning are land degradation and air pollution, as crop burning releases harmful greenhouse gases, higher levels of particulate matter and smog. It exacerbates loss of biodiversity and impacts human wellbeing and Land Surface Temperature (LST). It is important to understand the increase in LST on the areas surrounding the distributed CRB (Shurpali et al. 2019, Li et al. 2022, Lan et al. 2022).

The issue of landholding fragmentation impacts Indian agriculture. The land becomes divided among many inheritors if the same plot is passed down generation after generation. As a consequence, the amount of fragmented land rises and ongoing land reuse reduces soil fertility. It is also impractical to use intensive machinery to increase efficiency in these lands. By engaging sickles to physically chop the crop close to the ground, during the hand harvesting process, the crop residue is fed to the cattle. But when it comes to machine farming, about 40–50 cm of straw and 50–60 cm of crop stubble is produced which presents a problem for residue management (Mathur et al. 2019). Crop productivity, weather, crop type, and farming intensity affect how much crop residue is produced (Annual report by MNRE). Based on a number of prior researches, it can be said that burning crop residue is the primary cause of pollution in the environment, both locally and globally. Due to wind, the fog of gases emitted from burning crops travels further into the northern states of India and even crosses the Indian border.

There are very negligible debates about the managing of the waste produced by the various crops in the past studies or it could be said that the agricultural industry is not as regulated as that of municipal solid waste (MSW). Agri-waste is chiefly handled by the owners of the respective fields, which is a private sector with little involvement of the public sector (Bhuvaneshwari et al. 2019). Crop residue generation is not uniform all over the nation, and there exists a wide variation in the quantity of residue produced by the various regions. So, this is a vital topic among audiences beyond India for two reasons:

crop residue is an important integral of agriculture waste that can be used for the profit of society due to its organic composition and the second purpose is as discussed earlier in this section: the volume of crop residue is high and with unsustainable practices it has adverse effects on the environment beyond India (Bhuvaneshwari et al. 2019). In the current study, an effort is made to determine the effect of crop residue burning on Land Surface Temperature over a decade in the state of Punjab using LST product of MODIS to analyse the changes in the areas surrounding the burning locations.

Materials and Methods

Study area

Punjab is considered for the study as it is the chief state of India that burns crop residue. Punjab is a northern state that shares domestic boundaries with states of India – Haryana, Rajasthan, and Jammu and Kashmir, and an international boundary with Pakistan (Figure 1). It has a very fertile, alluvial plain with perennial rivers namely Sutlej, Ravi, Beas, Jhelum and Chenab. The average temperature during summer is around 40 to 47 degrees Celsius and in winter it is 5 to 12 degrees Celsius.

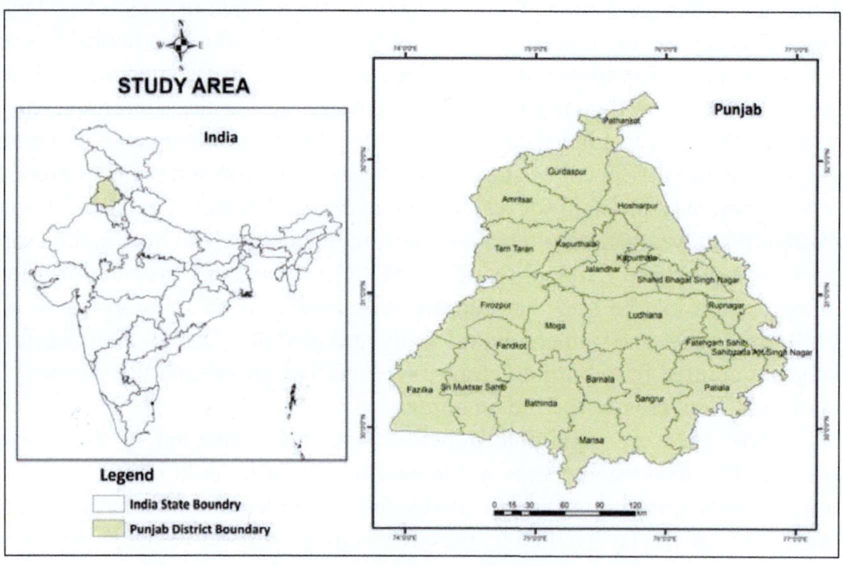

Figure 1. Punjab – study area.

Datasets used for the study

Monitoring crop residue burning using traditional methods is very difficult. Therefore, remote sensing as a technique is suitable for long term and large-scale observations. The obtained brightness temperature of the thermal infrared band can be used to monitor the local thermal anomalies. MODIS (Moderate-Resolution Imaging Spectroradiometer) is used for its ability to sense both smouldering and burning fires over 1000 m^2 in size. With good observing conditions, one-tenth of flaming fires could be detected and small fires at a resolution of 50 m^2 can also be detected in favourable observing conditions (Yan Zhuang et al. 2018, Zhang et al. 2020).

Methodology

For the current analysis the crop residue burning locations were extracted from MODIS and VIIRS fire product, and the LST from MODIS LST product. The fires were monitored by MODIS twice a day by the sensors Aqua and Terra, i.e., at night as well as during the day and 3.9 micrometre and 11 micrometre were used for this observation. From the overall points, the points that occur only in agriculture fields were retrieved for the present study.

Based on the burning sites the LST was extracted from the LST product to analyse the impact of CRB on local LST. Temperatures from non-burning sites were also extracted to show the difference between the surface temperatures of burnt sites as compared to non-burnt sites. ArcGIS software was majorly used for all this analysis. The point datasets of the burning sites along with temperature are incorporated in excel sheet and the variation over a period is done using descriptive statistics analysis for all the years mentioned. The LST is retrieved from MODIS LST data for a decade period and used to see the variation in temperature over the years and comparison is done with normal temperature of the region. The flow below chart is a representation of above description (Figure 2).

Once the LST were extracted based on the CRB locations the spatial distribution maps for all the years were prepared using interpolation technique in ArcGIS software.

Statistical method

In this study descriptive statistical method was used. The data was collected, organized, analysed and summarized in an understandable format such as graphs, tables and charts to make it computable. District wise as well as state level data was prepared and different graphs were prepared and mentioned in the result part of this study.

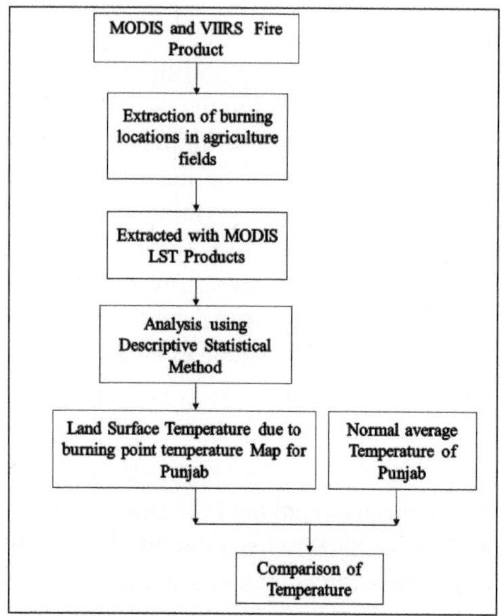

Figure 2. Methodology.

Results and Discussions

Yearly analysis of CRB on LST in Punjab State

Year wise from 2011–2020 for the whole of Punjab, data was analysed and temperature variations with respect to burnt and non-burnt sites along with the total number of burning sites were estimated as shown in Table 1. The

Table 1. Year=wise average normal temperature and burning temperature with burning sites.

Year	Burnt Temp. (°C)	NB Temp. (°C)	Difference (°C)	Burning Sites
2011	31.62	31.06	0.56	14591
2012	30.98	30.26	0.72	15859
2013	31.56	31.03	0.53	13463
2014	31.46	30.39	1.07	16051
2015	31.50	30.88	0.61	11861
2016	30.81	29.93	0.88	17424
2017	31.42	30.60	0.81	10996
2018	31.56	30.75	0.82	13306
2019	31.58	30.60	0.98	10308
2020	30.88	30.22	0.66	17376
Average Temperature	**31.33**	**30.57**	**0.76**	

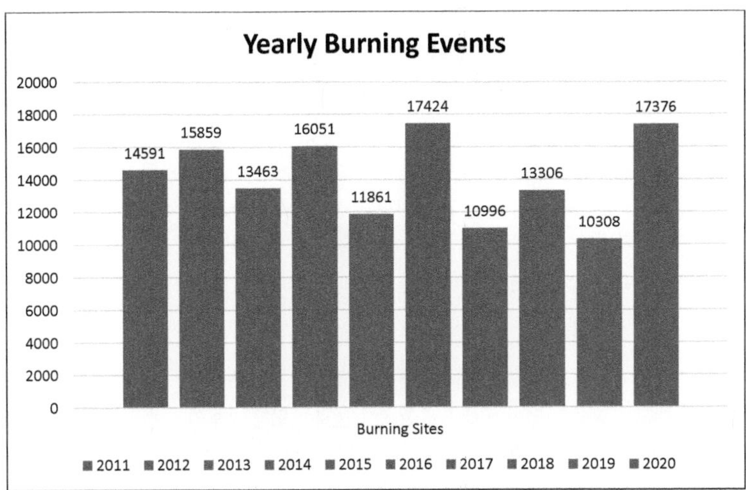

Figure 3. Graphical representation of the number of burning sites for every year.

study about the crop residue burning sites and its temperature variation over years was carried out using MODIS data. The burning sites are acquired from month of September to November for every year (Table 1) as well as for every district in the state of Punjab (Figure 1). The months were considered based on the cropping system of Kharif crops as mentioned in the introduction.

It can be seen from Table 1 that, maximum and minimum burnings sites occurred during the year 2016 and 2019 respectively. The number of burning sites of each year is graphically represented using a bar chart as shown in Figure 3. The average of the normal temperature of the region and LST increased due to burning are also derived and represented using graphical representation (Figure 4).

According to the analysis carried out on the basis of the information retrieved, it can be noted that the number of burning sites was maximum in the year 2016 with 17424 sites and minimum in 2019 with 10308 sites with an average burning temperature of 31.33°C. Throughout the years, the maximum burning sites occurred in the middle of October to the middle of November with LST ranging 30°C to 33°C. It is inferred that the number of burning sites doesn't depend on the burning temperature, but can depend on the quantity of the residue. The burning temperature of certain districts like Pathankot, Hoshiarpur, Amritsar, S.A.S. Nagar (Mohali), Mansa, Bathinda, and Faridkot show consistency over the period considered. From this it can be inferred that the crop residue burning activities has not reduced in these areas. The average temperature of the area has increased by 0.76°C in just over a decade, due to the burning of crop residues.

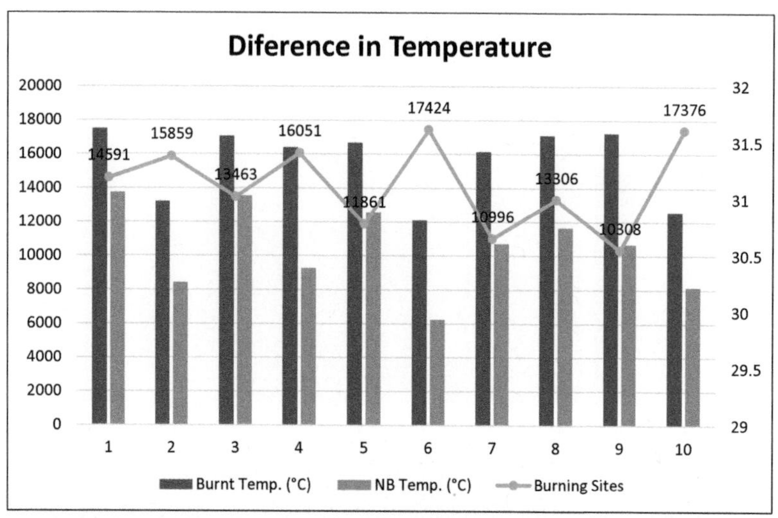

Figure 4. Representation of the burning sites with respective normal temperature and LST over the decade.

The spatial distribution maps of LST of Punjab for all the years are derived from the burning points using interpolation technique shown in Figure 5. From the image, it can be noted that the maximum value is around Fazilka, Muktsar, Bathinda and Mansa districts and minimum value regions are Pathankot, Gurdaspur, and Horisharpur districts of Punjab over all the years considered.

As discussed earlier, the rise in temperature may lead to change in climatic conditions in and around the region. The burning of residue has led to rise in temperature, due to release of harmful greenhouse gases which may have direct or indirect in global climatic conditions, and human health conditions. There are many case-studies available based on human health conditions in the crop burning areas especially in infants and old people. Several attempts were made by the Government of India as well as the State Government to educate the farmers about the advantages and drawbacks of crop residue burning through various government-initiated projects. There were many proposals to control the burning practices and encourage different practice methods for sustainable management. One of the policies introduced by the Ministry of Agriculture is the National Policy for Management of Crop Residue (NPMCR). But none of the schemes worked out in Punjab due to various *in situ* reasons.

The reasons are: (a) there is no diverse in crop rotation pattern and paddy remains the foremost crop which leads to large amount of residue; (b) the implementation of CRM (Crop Residue Management) machines like happy seeders and super seeders are minimum in number; and (c) many farmers complain that they can't afford them, (d) another major reason is the sentiment

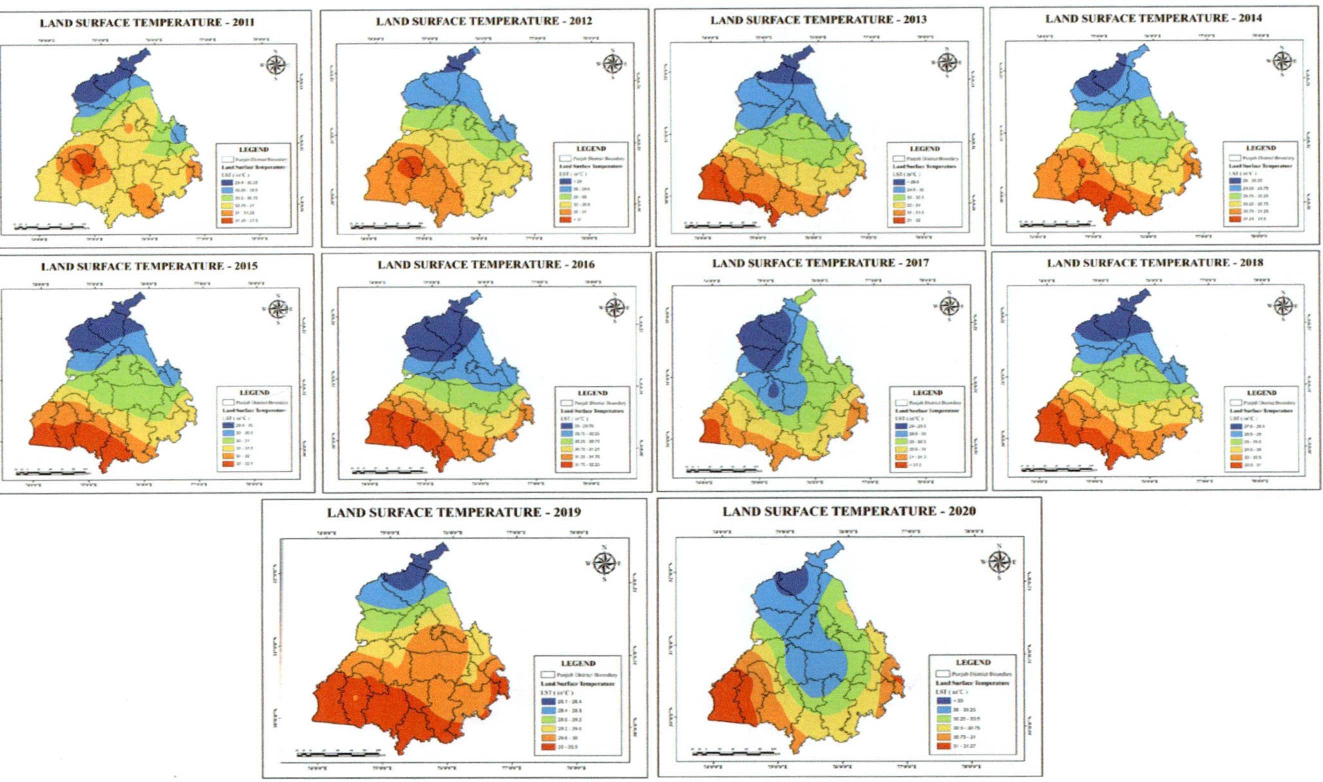

Figure 5. Spatial distribution maps of LST for Punjab from 2011 to 2020.

of farmers which acts as a barrier and does not let them accept change. In the year 2021, the state government took initiative by allowing industries to use crop residue in their boilers (Directorate of Information & Public Relation Punjab) which may be useful step in the coming years. Other than this, another method was also introduced: that was the PUSA decomposer which is considered eco-friendly and viable (Kurinji and Srish Prakash 2021).

Conclusions

The burning of crop residue in open is simple and cost effective, but not in the long run. Crop residue burning accelerates global warming and climate change. So, it is time to consider combating residue burning for good soil conditions, environmental safety, economic status and human health for the upcoming years using geospatial technologies. There are various ways to treat the crop residues namely in production of energy, fodder and bedding for animals, paper production, soil mulching, and mushroom cultivation. So, to initiate and empower the stakeholders, the state should launch various community programs for transportation of residue, equipment support, etc. to minimise the effect of burning and lessening the burden on the small-scale farmers and to create a path for sustainable development.

References

Annual Reports, Ministry of New and Renewable Energy, India. (n.d.). https://mnre.gov.in/file-manager/annual-report/2016-2017/EN/pdf/3.pdf.

Anuradha, Kuldeep Singh, Prashanth, C. S. and Raveena Bishnoi. 2021. Crop Residue burning and its harmful Impacts: A review. www.justagriculture.in. 2582–8223.

Bhuvaneshwari, S., Hiroshan Hetttiarachchi and Jay N. Meegoda. 2019. Crop residue burning in india: policy challenges and potential solutions. Int. J. Environ. Res. Public Health. 16: 832.

Kurinji, L. S. and Srish Prakash. 2021. Why Paddy Stubble Continues to be Burnt in Punjab. CEEW The Council.

Ritu Mathur and Srivastava, V. K. 2019. Crop Residue Burning: Effects on Environment.

Rong Li, Xinjie He, Hong Wang, Yi Wang, Meigen Zhang, Xin Mei et al. 2022. Estimating emissions from crop residue open burning in Central China from 2012 to 2020 using statistical models combined with satellite observations. Remote Sens. 14: 3682.

Ruoyu Lan, Sebastian D. Eastham, Tianjia Liu, Leslie K. Norford and Steven R. H. Barrett. 2022. Air quality impacts of crop residue burning in India and mitigation alternatives. Nature Communications 13: 6537.

Shurpali, N., Agarwal, A. K. and Srivastava, V. K. (eds.). 2019. Greenhouse Gas Emissions, Energy, Environment, and Sustainability. Springer Nature Singapore Pte Ltd.

Wenting Zhang, Mengmeng Yu, Qingqing He, Tianwei Wang, Lu Lin, Kai Cao et al. 2020. The spatial and temporal impact of agricultural crop residual burning on local land surface temperature in three provinces across China from 2015 to 2017. Journal of Cleaner Production 275: 124057.

Yan Zhuang, Ruiyuan Li, Hao Yang, Danlu Chen, Ziyue Chen, Bingbo Gao et al. 2018. Understanding temporal and spatial distribution of crop residue burning in China from 2003 to 2017 using MODIS data. Remote Sens. 10: 390.

6

Land Resource Mapping Framework for Delhi City using Urban Sprawl Modelling Methods

*Gaurav Kumar Mishra** and *Amit M. Deshmukh*

Introduction

The recommendations put forth in this study are aimed at informing the formulation of master plans and zoning policies, emphasizing the crucial role of understanding the intricacies of such plans. A master plan comprises a collection of documents, including a city map delineating various land uses through distinctive colors, transportation networks such as National Highways, State Highways, Primary and Secondary roads, streets, and railway lines. Additionally, it incorporates natural features like rivers and lakes, along with man-made structures such as canals and dams. The master plan also encompasses documents highlighting the city's socio-economic elements, providing demographic insights such as the male-to-female population ratio, adult-to-child population ratio, and the ratio of literate to illiterate population. While the master plan comprehensively addresses formal and coordinated development categories, there is limited information regarding the unplanned and uncoordinated growth that has transpired over

Department of Architecture and Planning, Visvesvaraya National Institute of Technology Nagpur, Maharashtra, India.

* Corresponding author: ergaurav.iitr@gmail.com; gaurav.mishra@students.vnit.ac.in

time. Consequently, our objective is to meticulously identify these instances of unplanned growth within each land-use category. Through our study, we aim to offer recommendations that not only assess the current scenario but also anticipate future developments, providing insights on how unplanned growth can be minimized and effectively managed in alignment with the overall urban development strategy.

After identifying areas with unplanned development, our focus shifts to assessing the suitability of these sites for potential redevelopment. To conduct this analysis, we plan to establish criteria based on insights gathered from questionnaire surveys conducted with the public. We aim to streamline these criteria by incorporating expert opinions and conducting thorough literature reviews. Once the criteria are refined, we intend to employ the Analytical Hierarchical Process, facilitating the systematic pair-wise comparison of criteria to determine the most suitable development sites. By ranking these sites, our findings can serve as valuable recommendations for policymakers, guiding them in selecting specific areas with higher developmental potential for strategic urban planning initiatives.

Several studies in the field have explored site suitability analysis for urban development, with notable work such as Anugya Shukla et al.'s examination of Lucknow city. In this study, the author delineated five sites based on land-use land-cover analysis and employed four criteria for Analytical Hierarchical Process to rank the most suitable sites. Other research endeavors also demonstrate the extensive use of GIS for preparing LULC maps and employing the AHP method for site suitability analysis. India is experiencing significant urbanization, with the United Nations projecting that half of the country's population will reside in cities by 2050. The primary driver of this urban expansion is rural-to-urban migration, which, despite challenges, contributes to economic growth. NITI AYOG reports indicate that by 2030, urban areas will account for 75% of the total GDP, designating cities as economic engines of the nation. The rapid and uncontrolled nature of urbanization poses a major challenge for city planners, necessitating proper infrastructure development to accommodate the burgeoning population. By 2050, approximately 500 million additional dwelling units will be required, underscoring the critical role of the government in providing infrastructure and land for urbanization. Assessing the current state of urbanization is crucial to understanding how formal and informal processes contribute to urban expansion. This information is essential for urban planners to minimize adverse effects on ecological and environmental aspects. Urban planners require tools to assess factors contributing to urban expansion, including the status, rate, pattern, and extent of sprawl. Without these tools, it is challenging to facilitate informed decision-making for the benefit of citizens (Ma and Xu 2010).

Remote sensing proves to be an invaluable tool in mapping spatial and temporal changes, offering insights into the variations in urban sprawl (Herold et al. 2003, Xiao et al. 2006). Analyzing urban sprawl historically provides users with valuable information for effective management and a comprehensive understanding of its evolution (Geymen and Baz 2007, Masek et al. 2000). Given that urban sprawl is intricately linked to land use and land cover changes, leveraging LULC change is essential for enhancing our comprehension of urban sprawl management.

To conduct quantitative analyses, various metrics are employed in literature. In India Sudhira et al. (2004) utilized Shannon entropy, patchiness, and built-up density, employing regression analysis to quantify urban sprawl in Udipi. In a study on Indore city, India; Kumar et al. (2007) combined landscape metrics and Shannon entropy. Similarly, Jat et al. (2008) applied Shannon entropy and landscape metrics to comprehend the sprawl dynamics in Ajmer city.

In addition to quantitative analyses, there is a need to evaluate qualitative measures for a comprehensive understanding of urban growth. Conducting a spatiotemporal assessment of urban expansion is crucial for identifying planning processes that have influenced urban growth. Such assessments enable planners to discern various types of urban developments and understand the planning interventions that have shaped urban expansion. The data derived from this type of assessment not only aids in recognizing regions experiencing rapid urbanization but also provides insights for policy formulation. Furthermore, it offers valuable information for assessing the environmental impact of urban growth and understanding the balance between planned and spontaneous development. Despite its importance, there is a limited body of research addressing both the qualitative and quantitative aspects of urban growth.

The focus of my research is on Delhi, India, the nation's capital, which has undergone significant urbanization in recent decades. This paper aims to provide a comprehensive analysis by incorporating both quantitative and qualitative measures. The study delves into the typology of development processes that have unfolded in the city over the past thirty-two years.

Study Area

Our research focuses on the National Capital Territory of Delhi, the esteemed national capital of India, situated in the northern part of the country. Encompassing an area of 1483 square kilometers, Delhi extends between 28.330 to 29.00 N latitude and 76.830 to 77.330 E longitude. According to the 2011 Census, Delhi is home to 11,034,555 residents, experiencing varying decadal growth rates, from 51.45% in 1981–91 to 47.02% in 1991–2001 and a notable decrease to 11.2% in 2001–2011. The population density in Delhi exhibited a range, from 9340 persons per km^2 in 2001 to 11320 persons per km^2 in 2011, marking the highest density among all states and union territories in India (Statistical Abstract of Delhi 2014). In economic terms, Delhi is poised for growth, with a projected annual GDP increase of 7.1% between 2016 and 2030, as outlined in the Global Cities 2030 report (2016). This comprehensive analysis provides a foundation for understanding the demographic and economic dynamics of Delhi.

Datasets

These datasets offer valuable contributions due to their medium resolution, extensive archival history, and consistent spectral and radiometric resolutions. Notably, Landsat images stand out as reliable and freely accessible sources, with the first imagery obtained from this satellite in 1972. A significant enhancement occurred in 1982 with the addition of the Thematic Mapper instrument, providing seven spectral bands and a spatial resolution of 30 meters. This upgrade proved highly beneficial for scrutinizing urban growth patterns. Consequently, our study leveraged Landsat images from the years 1989, 1994, 1999, 2004, 2009, and 2014 to comprehensively evaluate the urban expansion of Delhi. To complement this satellite data, toposheets specific to the study area were obtained from the Survey of India. These toposheets are projected onto the Universal Transverse Mercator (UTM) coordinate system, utilizing the WGS 1984 datum as a reference. In the preprocessing phase, the images underwent mosaicking and clipping processes to precisely define the study area. The Survey of India toposheets played a crucial role as ancillary data, employed both in the classification process and for accuracy assessment. Additionally, these toposheets provided essential details regarding the locations of national and state highways, serving as a foundation for the creation of a dedicated layer in our analysis. This integration of diverse datasets ensures a comprehensive and accurate exploration of urban expansion dynamics in Delhi.

The Landsat imageries encompass various layers, ranging from band 1 to band 8, each representing distinct electromagnetic wavelengths. These wavelengths span the visible spectrum, including red, green, and blue; the ultraviolet range; and the infrared range, which includes both thermal and near-infrared components. Through layer stacking, compositions like false-color composites are generated, offering different combinations of three channels: red, green, and blue. Specific combinations serve various purposes, such as extracting information about vegetation through a false-color composite of satellite imageries. To derive information pertinent to the urban built-up class or to collect training samples for supervised classification, different channel combinations are employed. This classification process ultimately yields a comprehensive land-use land-cover map. Thematic datasets for the year 2019 were prepared using Landsat imageries and processed within a GIS environment, utilizing ArcGIS software. These thematic datasets encompass parameters such as proximity to roads, distance to the Central Business District (CBD), population density, and slope values. The delineation of government-restricted areas was accomplished through the analysis of Delhi's proposed land-use plan using ArcGIS software, as illustrated in Figure 3. Furthermore, population densities for various years were computed by considering different zones within Delhi, a task executed through ArcGIS software. Python programming was employed to process all these layers during model training and predicting future urban built-up expansion. This integrated approach, involving both remote sensing and GIS techniques, ensures a robust foundation for understanding and forecasting urban dynamics.

Methods

A brief method has been described through Figure 1.

Image processing and classification

The research was conducted across a land area spanning 1483 square kilometers. To enhance data accuracy, an empirical image-based Quick Atmospheric Correction (QUAC) method was employed for atmospheric correction of the

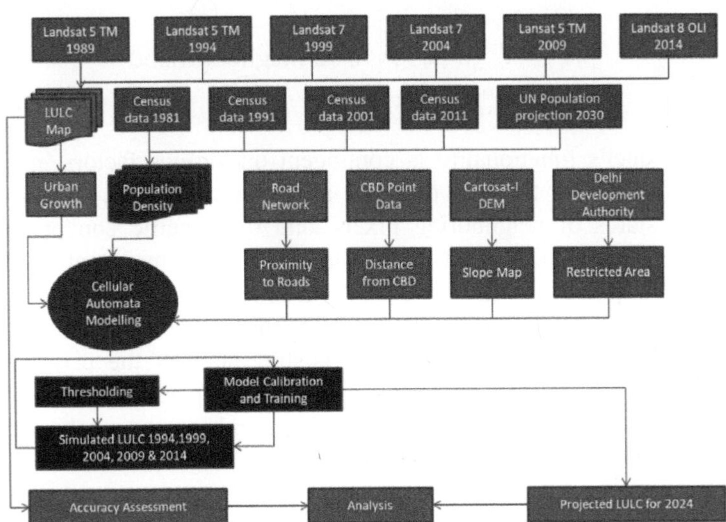

Figure 1. Flow chart showing LULC projection methodology.

acquired imagery. Subsequently, a maximum likelihood classifier was applied to perform supervised classification on the atmospherically corrected imagery. The classification involved categorizing each Landsat image into four distinct classes: Built-up, Forest, Water, and Other.

LULC (Land-use Land-cover) change detection

To comprehend the pattern of urban growth, it becomes imperative to analyze changes in land-use and land cover. Remote sensing emerges as a preferred tool for such analysis due to its efficiency in terms of time, cost, and operational simplicity (Jensen and Im 2007). In Figure 2, the land-use land cover (LULC) maps are presented alongside auxiliary maps, offering a comprehensive visual representation of the evolving landscape. This method proves instrumental in gaining insights into the dynamic nature of urban development.

CA Model Algorithm

This algorithm, employed for forecasting urban growth in Delhi, takes into account a comprehensive set of factors influencing urban expansion. Utilizing a 3×3 size kernel, as depicted in equation (i), ensures a thorough examination of the spatial dynamics and interactions among various contributing elements. This approach is crucial to capture the complexity of urban growth patterns and enhance the accuracy of predictions. The algorithm's consideration of multiple factors makes it a robust tool for simulating and understanding the intricate processes associated with urban development in the study area.

$$A_{i,j}^{(t)} = \begin{bmatrix} a_{i-1,j-1}^{(t)} a_{i-1,j}^{(t)} a_{i-1,j+1}^{(t)} \\ a_{i,j-1}^{(t)} a_{i,j}^{(t)} a_{i,j+1}^{(t)} \\ a_{i+1,j-1}^{(t)} a_{i+1,j}^{(t)} a_{i+1,j+1}^{(t)} \end{bmatrix} \quad 3 \times 3 \text{ neighbourhood} \tag{1}$$

This model's functionality is contingent on various factors, including the classification of the test pixel as urban or non-urban, the urban or non-urban status of neighboring pixels, and the specific conversion rule applied for the transition from non-urban to urban (Kumar et al. 2009). To illustrate, if a pixel is in its current state at time t, its state at time t + 1 is determined by the predefined rules governing the conversion from non-urban to urban (ϕ), in conjunction with the existing state of the pixel (Kumar et al. 2009). This dynamic process incorporates temporal dependencies and rule-based transformations, offering a comprehensive mechanism for simulating urbanization patterns over time.

$$a_{i,j}^{t+1} = \phi(A_{i,j}^{t}) \tag{2}$$

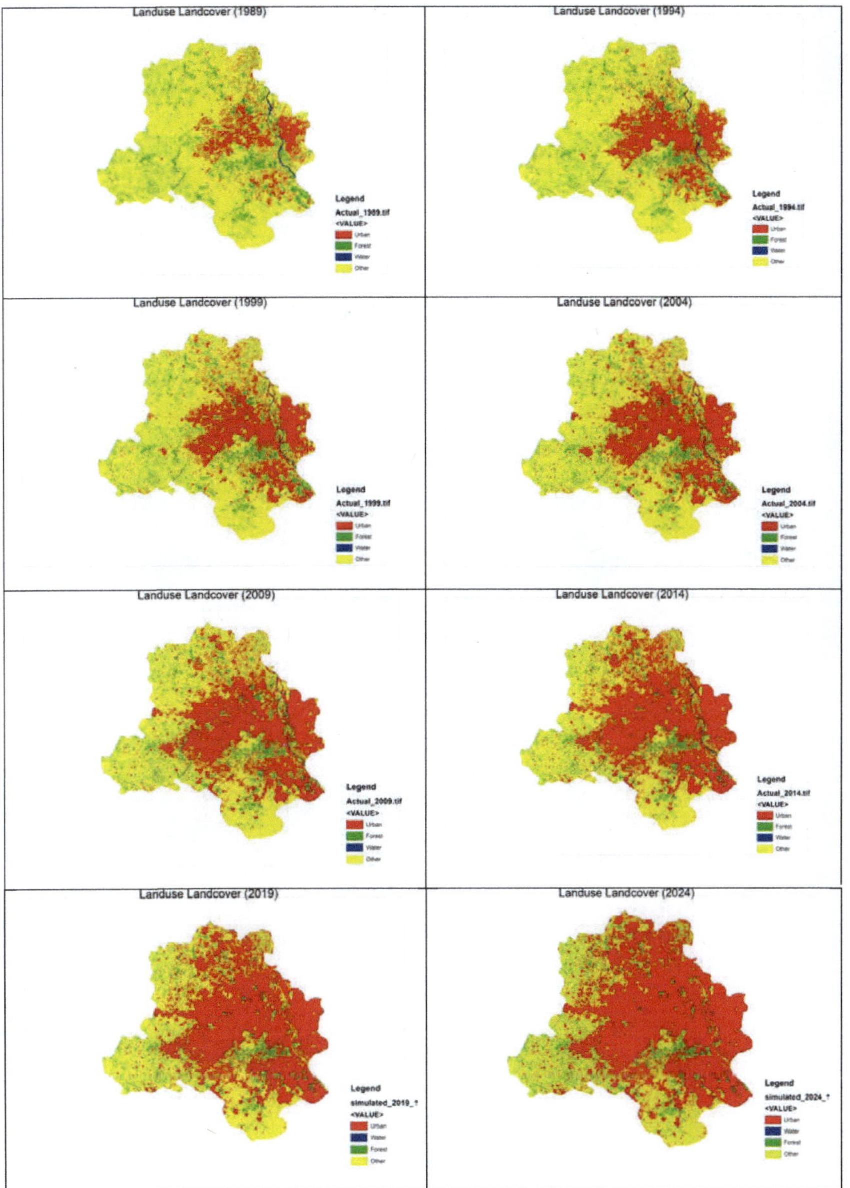

Figure 2 contd. ...

...Figure 2 contd.

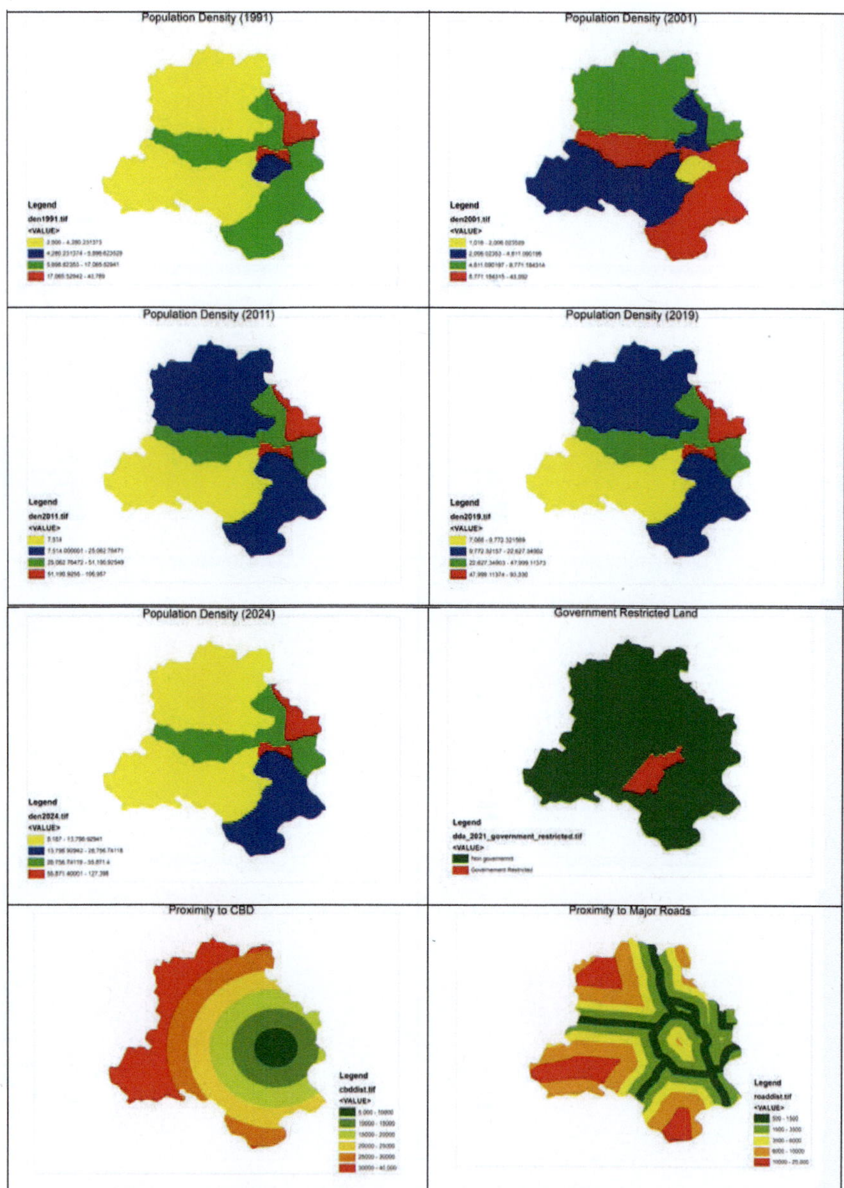

Figure 2. Data Layers used in Geospatial Modelling.

The pixel transition from non-urban to urban (ϕ) is governed by a set of conditional statements embedded with threshold values. These rules dictate the specific conditions under which a pixel undergoes the transformation, introducing a nuanced and parameterized approach to the simulation process. The threshold values act as benchmarks, determining the extent to which the pixel characteristics align with the predefined conditions for urbanization. This rule-based framework enhances the model's adaptability and allows for fine-tuned adjustments based on diverse criteria, contributing to a more realistic representation of urban growth dynamics.

$$\phi = f(T, B) \tag{3}$$

In this equation, the rules governing the conversion from non-urban to urban are expressed as a function of two key components: T and B. T represents the set of threshold values corresponding to various factors influencing urban growth, encompassing a comprehensive range of parameters. These thresholds act as critical benchmarks, guiding the transformation process based on specific conditions related to each factor. On the other hand, B constitutes a set of urban count values within the kernel, intricately linked to each set of threshold values in T. This interplay between threshold values and urban count values introduces a nuanced and context-specific dimension to the model, allowing for a detailed consideration of multiple factors influencing the urbanization process.

$$T = \{T_R + T_C + T_P + T_S\} \tag{4}$$

$$B = \{B_R + B_C + B_P + B_S\} \tag{5}$$

In this formulation, the variables T_R, T_C, T_P, and T_S represent the threshold values corresponding to factors such as proximity to roads, distance from CBD, population density, and slope value, respectively. Similarly, B_R, B_C, B_P, and B_S signify the associated counts of urban built-up pixels within the test kernel for each respective factor. The conversion rules for transitioning from non-urban built-up pixels to urban built-up pixels are grounded in real-world conditions. Two fundamental rules govern this conversion process. Firstly, the classes representing urban built-up land and water bodies are exempted from elimination. Secondly, the class pertaining to vegetation land or others has the potential to transform into urban land, contingent on meeting specific criteria defined by threshold values (T) and the count of neighboring urban built-up pixels (B). Notably, this conversion is permissible only if the test pixel falls within the non-restricted category in the restricted areas layer, emphasizing the influence of geographic constraints on urban expansion.

Model Calibration

To calibrate the model, Land Use Land Cover (LULC) maps derived from Landsat Satellite imagery were utilized. The simulation process involved using the LULC map of 1989 to simulate the map for 1994, followed by using the 1994 map to simulate the 1999 map, and so forth. The final simulation for the year 2014 was accomplished by using the 2009 LULC map. This sequential approach facilitated the determination of optimal threshold values, ensuring the simulated maps closely mirrored the real-world scenarios (Kumar et al. 2009). The calibration process employed a statistical and spatial simulation for both LULC maps and maps generated for other influential parameters. Trial and error played a pivotal role in obtaining threshold values for four factors: T_R (Threshold Rainfall), T_C (Threshold Temperature), T_P (Threshold Population), and T_S (Threshold Slope). The associated pixel count values for built-up land (BR, BC, BP, and BS) were also derived through this iterative process. Notably, the simulation monitored the generation of new built-up pixels influenced by each contributing factor. Post-calibration, the established threshold values were projected into the future to forecast the extent of future built-up areas. The accuracy of each simulation step was assessed using Principal Component Analysis (PCA) (Kumar et al. 2009). Incorporating the spatial differences between two images, the PCA involved subtracting the built-up pixels of time t2 from those of time t1. Built-up layers were prepared from Landsat Satellite imagery, and the simulated LULC map for the corresponding period. The subtraction process revealed non-zero values (1 or −1) for pixels where there was no match, indicating areas of change or growth. This comprehensive approach ensured the reliability and accuracy of the model throughout its various stages.

Results

LULC (Land-use Land-cover) classification

The evaluation of the classified image's accuracy was conducted by comparing it with ground truth data. Ground truth data provides a reference point to assess the reliability and precision of the classification results. In this context, various metrics and statistical measures were employed to quantify the agreement between the classified image and the actual ground conditions. This rigorous accuracy assessment not only validates the robustness of the classification process but also ensures the credibility of the derived land-use land-cover map. Additionally, field surveys and verification procedures were implemented to enhance the reliability of the ground truth data, contributing to a comprehensive and trustworthy assessment of the classified image.

LULC (Land-use Land-cover) projection for the year 2024

The comprehensive utilization of the sequential Landsat datasets from 1989 to 2019 formed the basis for projecting the Land Use Land Cover (LULC) scenario for the year 2024, as illustrated in Figure 2. The classified images corresponding to each year were procured to capture the dynamic changes in land cover over the decades. In conjunction with these images, additional datasets such as population density records for each year, identification of major government restricted areas, proximity to key road networks, and distance from the Central Business District (CBD) were incorporated. The preparation and analysis of these diverse datasets were accomplished using ArcGIS software, and the projection of the dataset into the future, specifically for 2024, was executed through the application of the Cellular Automata Algorithm in Python. This integrative approach ensures a robust and forward-looking representation of the anticipated LULC patterns, considering both historical trends and influencing factors.

Analysis of Urban Expansion

Urban landscape has undergone a substantial transformation, expanding from a modest 161 square kilometers in 1989 to a sprawling 785 square kilometers by the year 2024. This remarkable growth signifies a dynamic shift in the city's spatial configuration over the years. The expansion has not only reshaped the physical boundaries of the urban area but has also brought about significant changes in the socio-economic and infrastructural aspects of the region. The evolving urban expanse underscores the need for proactive urban planning strategies to ensure sustainable development and effective management of resources in the face of continuous expansion.

Table 1 helps us analyze that urban expansion had taken place in the year 1994 from 1989 and the difference between the urban area was 110 square kilometers during the intervening years. Forest area got saturated by a quantity of 5 square kilometers in 1994 in comparison with the area

Table 1. LULC change matrix for 1989–1994.

Land use Landcover Change Matrix (1989–1994)					
1994 Assessment	Urban	Forest	Water	Other	Total 1989
Urban	161	5	1	104	161
Forest	5	198	0	0	198
Water	1	0	13	0	13
Other	104	0	0	1129	1129
Total 1994	271	193	12	1025	1501
Net change	+110	−5	−1	−104	

of forest in the year 1989. Similarly, water bodies got saturated by 1 square kilometers in the year 1994 in comparison with the year 1989. Other land uses were also taken by urban area and 104 square kilometers saturation in other land uses were found in the year 1994 comparison with the year 1989. So, we can say here that only urban area has been found to have expanded and it has been also found that the areas of expansion were encroached in the forest, water, and other land uses. So, it becomes essential to find the historically evolution of urban areas so that it can be manageable in upcoming years because other land covers also play important role in maintaining the quality of life for living beings.

From Table 2 we can analyze that urban expansion had taken place in the year 1999 from 1994 and the difference between the urban area was 80 square kilometers during the intervening period. Forest area got saturated by quantity of 2 square kilometers in 1999 in comparison with the area of forest in the year 1994. Similarly, water bodies got saturated by 4 square kilometers in the year 1999 in comparison with the year 1994. Other land uses were also taken by urban area and 74 square kilometers saturation in other land uses were found in the year 1999 comparison with the year 1994. So, we can say here that only urban area has been found with its expansion and it has been also found that the areas of expansion were encroached in the forest, water, and other land uses. So, it becomes essential to find the historically evolution of urban areas so that it can be manageable in upcoming years because other land covers also play important role in maintaining the quality of life for living beings.

From Table 3 we can analyze that urban expansion had taken place in the year 2004 from 1999 and the difference between the urban area was 73 square kilometers in the intervening period. Forest area got saturated by quantity of 5 square kilometers in 2004 in comparison with the area of forest in the year 1999. Similarly, water bodies got saturated by 0.5 square kilometers in the year 2004 in comparison with the year 1999. Other land uses were also taken by urban area and 67 square kilometers saturation in other land uses were

Table 2. LULC change matrix for 1994–1999.

Land use Landcover Change Matrix (1994–1999)					
1999 Assessment	Urban	Forest	Water	Other	Total 1994
Urban	271	2	4	74	271
Forest	2	193	0	0	193
Water	4	0	12	0	12
Other	74	0	0	1025	1025
Total 1999	351	191	8	951	1491
Net change	+80	–2	–4	–74	

Table 3. LULC classes change matrix for 1999–2004.

Land use Landcover Change Matrix (1999–2004)					
2004 Assessment	Urban	Forest	Water	Other	Total 1999
Urban	351	5	0	67	351
Forest	5	191	0	0	191
Water	0	0	8	0	8
Other	67	0	0	951	951
Total 2004	424	186	8	884	1501
Net change	+73	−5	± 0	−67	

Table 4. LULC change matrix for 2004–2009.

Land use Landcover Change Matrix (2004-2009)					
2009 Assessment	Urban	Forest	Water	Other	Total 2004
Urban	424	10	0	84	424
Forest	10	186	0	0	186
Water	0	0	8	3	8
Other	84	0	3	884	884
Total 2009	518	176	11	797	1502
Net change	+94	−10	+3	−87	

found in the year 2004 comparison with the year 1999. So, we can say here that only urban area has been found with its expansion and it has been also found that the areas of expansion were encroached in the forest, water, and other land uses. So, it becomes essential to find the historically evolution of urban areas so that it can be manageable in upcoming years because other land covers also play important role in maintaining the quality of life for living beings.

From Table 4 we can analyze that urban expansion had taken place in the year 2009 from 2004 and the difference between the urban area was 94 square kilometers in the intervening period. Forest area got saturated by a quantity of 10 square kilometers in 2009 in comparison with the area of forest in the year 2004. Similarly, water bodies increased by 3 square kilometers in the year 2009 in comparison with the year 2004. Other land uses were also taken by urban area and 87 square kilometers saturation in other land uses were found in the year 2009 comparison with the year 2004. So, we can say here that only urban area has been found with its expansion and it has been also found that the areas of expansion were encroached in the forest, and other land uses. So, it becomes essential to find the historically evolution of urban areas so that it can be manageable in upcoming years because other land covers also play important role in maintaining the quality of life for living beings.

From Table 5 we can analyze that urban expansion had taken place in the year 2014 from 2009 and the difference between the urban area was 104 square kilometers in the intervening period. Forest area got saturated by a quantity of 14 square kilometers in 2014 in comparison with the area of forest in the year 2009. Similarly, water bodies got saturated by 4 square kilometers in the year 2014 in comparison with the year 2009. Other land uses were also taken by urban area and 86 square kilometers saturation in other land uses were found in the year 2014 in comparison with the year 2009. So, we can say here that only urban area has been found to have been expanded and it has been also found that the areas of expansion were encroached in the forest, water, and other land uses. So, it becomes essential to find the historically evolution of urban areas so that it can be manageable in upcoming years because other land covers also play important role in maintaining the quality of life for living beings.

From Table 6 we can analyze that urban expansion had taken place in the year 2019 from 2014 and the difference between the urban area was 84 square kilometers during the intervening period. Forest area got saturated by a quantity of 12 square kilometers in 2019 in comparison with the area of forest in the year 2014. Similarly, water bodies got saturated by 0.5 square

Table 5. LULC classes change matrix for 2009–2014.

Land use Landcover Change Matrix (2009–2014)					
2014 Assessment	Urban	Forest	Water	Other	Total 2009
Urban	518	14	4	86	518
Forest	14	176	0	0	176
Water	4	0	11	0	11
Other	86	0	0	797	797
Total 2014	622	162	7	711	1502
Net change	+104	–14	–4	–86	

Table 6. LULC classes change matrix for 2014–2019.

Land use Landcover Change Matrix (2014–2019)					
2019 Assessment	Urban	Forest	Water	Other	Total 2014
Urban	622	12	0	72	622
Forest	12	162	0	0	162
Water	0	0	7	0	7
Other	72	0	0	711	711
Total 2019	706	150	7	639	1502
Net change	+84	–12	±0	–72	

Table 7. LULC classes change matrix for 2019–2024.

Land use Landcover Change Matrix (2019–2024)					
2024 Assessment	Urban	Forest	Water	Other	Total 2019
Urban	706	14	0	65	706
Forest	14	150	0	0	150
Water	0	0	7	0	7
Other	65	0	0	639	639
Total 2024	785	136	7	574	1502
Net change	+79	−14	±0	−65	

kilometers in the year 2019 in comparison with the year 2014. Other land uses were also taken by urban area and 72 square kilometers saturation in other land uses were found in the year 2019 comparison with the year 2014. So, we can say here that only urban area has been found to have been expanded and it has been also found that the areas of expansion were encroached in the forest, water, and other land uses. So, it becomes essential to find the historically evolution of urban areas so that it can be manageable in upcoming years because other land covers also play important role in maintaining the quality of life for living beings.

From Table 7 we can analyze that urban expansion will take place in the year 2024 from 2019 and the difference between the urban area will be 79 square kilometers in this duration. Forest area will be saturated by a quantity of 14 square kilometers in 2024 in comparison with the area of forest in the year 2019. Similarly, water bodies will be saturated by 0.5 square kilometers in the year 2024 in comparison with the year 2019. Other land uses will also be taken by urban area and 65 square kilometers saturation in other land uses will be found in the year 2024 comparison with the year 2019. So, we can say here that only urban area has been found to have expanded and it has been also found that the areas of expansion were encroached in the forest, water, and other land uses. So, it becomes essential to find the historically evolution of urban areas so that it can be manageable in upcoming years because other land covers also play important role in maintaining the quality of life for living beings.

Figure 3 provides a clear depiction of the substantial growth in urban land use, exhibiting a remarkable expansion rate. The encroachment of urban expansion on other land uses is evident, with the projected urban area reaching approximately 800 square kilometers by 2024. Consequently, the areas designated for forest, water, and other land uses are expected to experience a reduction in their respective spatial extents. This trend underscores the transformative impact of urbanization on the landscape and the subsequent dominance of urban land use in the study area.

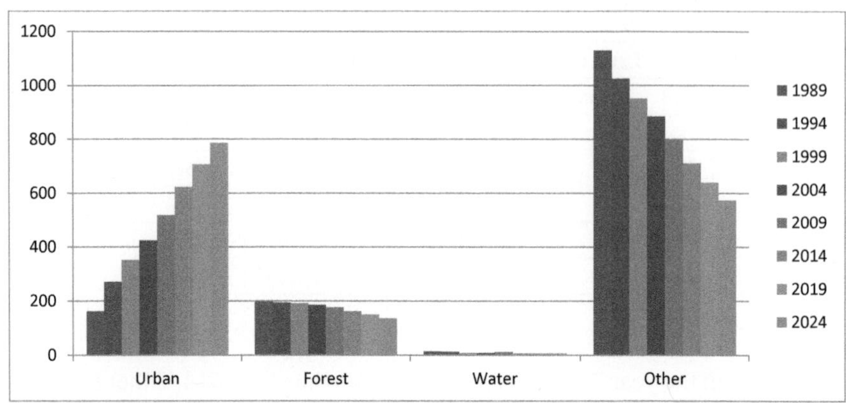

Figure 3. LULC classes area in square meters for each interval of the study period.

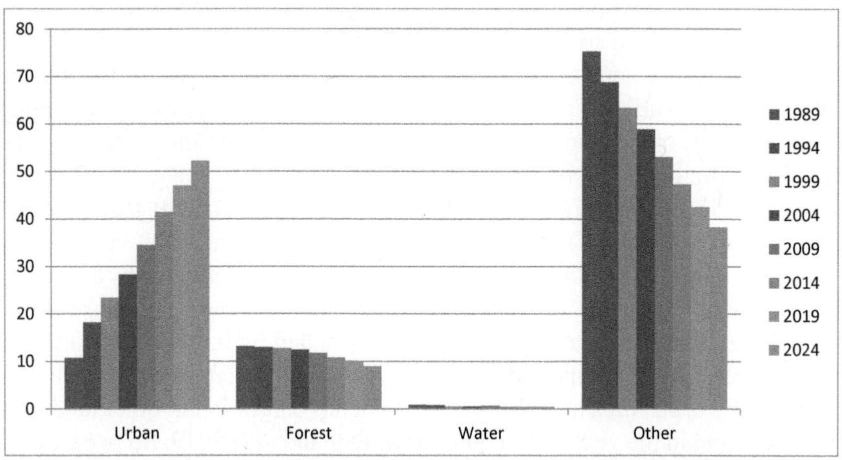

Figure 4. Percentage-wise distribution of LULC classes for each interval of study period.

In Figure 4 the proportional distribution of land use categories in the National Capital Territory (NCT) of Delhi is illustrated. Notably, the urban area's percentage share has surged significantly, escalating from 10% in 1989 to a substantial 53% by the year 2024. Meanwhile, land uses categorized as forest, water, and others exhibit comparatively lower percentages in terms of their spatial coverage in the year 2024. This representation underscores the transformative shift in land use dynamics, particularly the remarkable expansion of urban areas at the expense of the other class.

Figure 5 illustrates the alterations observed in various classes of land use and Land cover (LULC). This visual representation captures the dynamic shifts and modifications that have occurred over time in different land use categories. Analyzing the changes in these LULC classes provides

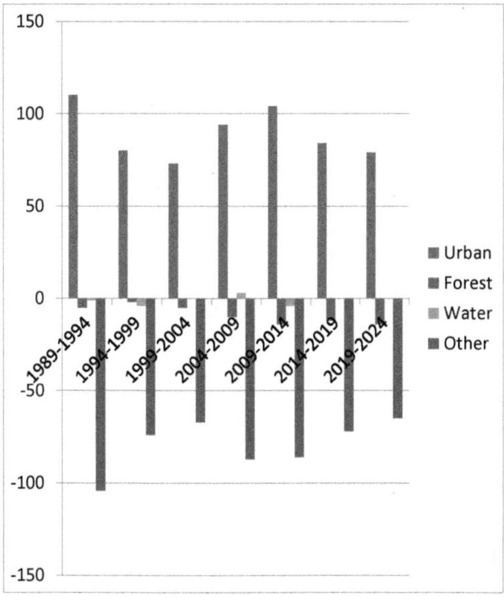

Figure 5. Net change in the LULC classes in succeeding years from 1989 to 2024 with the interval of 5 years.

valuable insights into the evolving landscape, showcasing transformations in urban, forest, water, and other land uses. The visual depiction serves as a comprehensive overview of the fluctuating patterns within the studied area, aiding in a nuanced understanding of the broader land use dynamics.

Figure 6 presents a comprehensive breakdown of the percentage distribution for each land use class across multiple years in the dataset. This visual representation offers a detailed and year-specific portrayal of the proportionate contributions of urban, forest, water, and other land uses. Examining these percentages over the temporal scope of the dataset provides

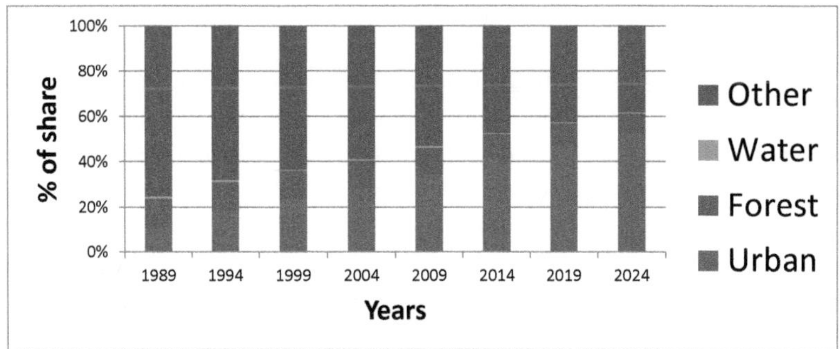

Figure 6. Percentage share of individual LULC class in the study area from year 1989 to 2024 with interval of 5 years.

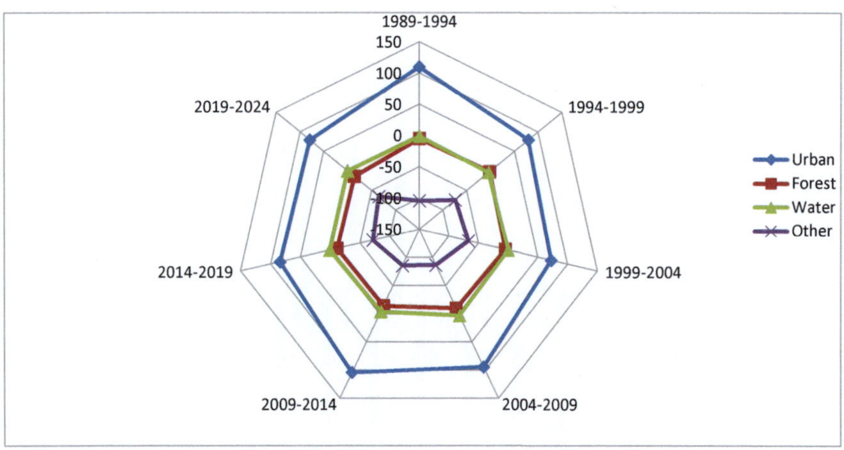

Figure 7. Depiction of LULC classes through a radar diagram.

a nuanced understanding of the relative dominance and changes within each land use category throughout the studied period. This graphical representation aids in discerning the evolving trends and proportional variations, contributing to a more in-depth analysis of the dataset's temporal dynamics.

Figure 7 employs a radar chart to illustrate the change pattern and quantity of land use and land cover (LULC) class changes. The distinctive feature of a regular heptagon in this representation suggests a proportional growth pattern among the different years under consideration. The geometry of the radar chart provides a visual tool for assessing the consistency and balance in the transformation of LULC classes across the temporal spectrum. The regularity in the heptagon shape indicates a systematic and proportionate evolution in LULC dynamics over the specified time intervals. This graphical representation aids in identifying trends and deviations, facilitating a qualitative understanding of the dataset's transformational aspects.

Conclusion

In conclusion, the escalating demand for urban areas is progressively encroaching upon other land uses, as evidenced by the spatial analysis conducted using time series data and employing Python and ArcGIS software. The study has successfully justified the spatial location of urban expansion, considering key constraints such as government-restricted areas and slope. By incorporating additional constraints, such as environmental protection measures, for instance, using forest area, the model for urban expansion can be further refined. The research not only explores the dynamics of urbanization in the past but also provides insights into future urbanization trends. The findings emphasize the need for a sustainable approach to urban expansion, given the anticipated substantial increase. Proper allocation of this urban expansion is crucial, considering the limited land resources, and underscores the importance of using land resources judiciously and sustainably. This study contributes to a better understanding of urbanization processes, highlighting the interactions among various factors shaping urban development.

Future Scope

This modelling framework can be used for multi-scenario simulations where economic, social, and environmental factor can act.

References

Geymen, A. and Baz, I. 2007. Monitoring urban growth and detecting land-cover changes on the Istanbul metropolitan area. Environmental Monitoring and Assessment.

Herold, M., Goldstein, N. C. and Clarke, K. C. 2003. The spatiotemporal form of urban growth: measurement, analysis and modeling. Remote Sensing of Environment.

Jat, M. K., Garg, P. K. and Khare, D. 2008. Modelling of urban growth using spatial analysis techniques: a case study of Ajmer city (India). International Journal of Remote Sensing.

Jensen, J. R. and Im, J. 2007. Remote sensing change detection in urban environments. pp. 7–31. *In*: Jensen, R. R., Gatrell, J. D. and McLean, D. (eds.) Geo-Spatial Technologies in Urban Environments. Heidelberg: Springer Berlin, Heidelberg.

Kumar, J. A., Pathan, S. K. and Bhanderi, R. J. 2007. Spatio-temporal analysis for monitoring urban growth—a case study of Indore City. Journal of the Indian Society of Remote Sensing.

Kumar, U., Mukhopadhyay, C. and Ramachandra, T. V. 2009. Cellular automata and Genetic Algorithms based urban growth visualization for appropriate land use policies. Proceedings of the Fourth Annual International Conference on Public Policy and Management. Bangalore: IIM B.

Ma, Y. and Xu, R. 2010. Remote sensing monitoring and driving force analysis of urban expansion in Guangzhou City, China. Habitat International.

Masek, J. G., Lindsay, F. E. and Goward, S. N. 2000. Dynamics of urban growth in the Washington DC metropolitan area, 1973–1996, from Landsat observations. International Journal of Remote Sensing.

Nations, U. 2016. Global Cities 2030 Report. United Nations.

Shukla, A. and Jain, K. 2018. Modeling urban growth trajectories and spatiotemporal pattern: a case study of Lucknow City, India. Journal of the Indian Society of Remote Sensing.

Statistics, D. O. 2014. Statistical Abstract of Delhi. New Delhi: Directorate of Economics & Statistics.

Sudhira, H., Ramachandra, T. and Jagadish, K. 2004. Urban sprawl: metrics, dynamics and modelling using GIS. International Journal of Applied Earth Observation and Geoinformation.

Xiao, J., Shen, Y., Ge, J., Tateishi, R., Tang, C., Liang, Y. et al. 2006. Evaluating urban expansion and land use change in Shijiazhuang, China, by using GIS and remote sensing. Landscape and Urban Planning.

7

Environmental Intelligence
Mapping the Transforming Landscape through Artificial Intelligence and Satellite Data

Chatrabhuj[1],* and *Kundan Meshram*[2]

Introduction

In recent decades, there has been a remarkable surge in global urbanization. While urbanization is a prevalent trend worldwide, it has gained even greater momentum in developing nations due to their rapid economic expansion. India currently hosts approximately 34% of its total population in cities, and this figure is expected to substantially increase in upcoming years (Liu et al. 2023). With the nation's current urbanization rate standing at 1.1%, it is on track to become the most populated urban center globally by 2050, boasting an estimate that the urban population will rise in the near future (Arowolo and Deng 2018). To accommodate this, cities are developing well beyond their traditional urban boundaries, placing significant strain on surrounding natural resources. This expansion results in the displacement of critical lands such as marshlands, forests, wet areas, and shallow water bodies. Such rapid urban growth is a key factor in a multitude of environmental challenges, including

[1] Research Scholar, Department of Civil Engineering, Guru Ghasidas Vishwavidyalaya, Bilaspur-495009, Chhattisgarh, India.

[2] Assistant Professor, Department of Civil Engineering, Guru Ghasidas Vishwavidyalaya, Bilaspur-495009, Chhattisgarh, India.

* Corresponding author: chatrabhuj513@gmail.com

climate change, degradation of ecosystem services, urban heat islands, heatwaves, flash floods, loss of fertile land, an increment in urban areas, and it leads to change in the temperature of cities.

A prominent outcome of Land Use and Land Cover (LULC) modifications within urban environments is the change in Land Surface Temperature (LST), contributing to the formation of Urban Heat. LST serves as a crucial parameter linked to climate changes in cities (Guha et al. 2022). These variable experiences significant fluctuations in response to land pattern cover configurations, making it clear that alterations in land conversion significantly influence land surface temperatures (Roy and Bari 2022).

The existing literature highlights that the multispectral bands of thermal infrared bands generated from various Landsat image sensors have significantly improved the capabilities of remote sensing techniques to simultaneously create LST and LULC information using the same dataset (Martins et al. 2014). An important edge of Landsat data is its ability to cover temporal data, spanning several decades, in contrast to MODIS and ASTER (Prishchepov et al. 2012). The broader coverage of Landsat data enables conducting seasonal and temporal analyses at much higher spatial resolutions, providing a valuable resource for long-term monitoring and assessment.

Using high-resolution satellite imagery to estimate low-performing areas in rural areas of China region sets (Wu and Zhang 2012). The study showed that satellite imagery-based poverty mapping could provide accurate and timely data for poverty monitoring, which can aid in poverty reduction efforts. Similarly (Fu and Weng 2016) used satellite imagery to monitor agricultural production in Ukraine. The study demonstrated that remote sensing techniques could provide data on crop yield and land use changes, which are critical for sustainable agriculture monitoring and drinking water source protection (Wang et al. 2020).

Remote sensing techniques, particularly using satellite imagery, offer a valuable and effective means of monitoring and evaluating progress towards Sustainable Development Goals (SDGs). These techniques provide spatially explicit information about various factors like forest cover, area under agriculture, water bodies area, urban land, and infrastructure, which are crucial for tracking SDG-related indicators (Haladar et al. 2023, Mu et al. 2019) Research has demonstrated their utility in applications like poverty mapping, agricultural production monitoring, and even energy balance observations from moon-based platforms. These methods can significantly contribute to achieving global development targets and guiding informed policy decisions in areas like poverty reduction and sustainable agriculture.

In summary, the utilization of RS and GIS techniques combined has proven to be invaluable in monitoring and understanding LULC changes in various areas (Seyam et al. 2023). These technologies have enabled researchers to track changes in urban built-up areas, individual tree recognition, and agricultural land use, among other applications. The results obtained through these methods contribute to better spatial planning, environmental conservation, and sustainable development efforts (Wang et al. 2020). Whether it is improving nighttime light data for urban extraction, developing advanced frameworks for tree detection, or predicting land use patterns for sustainability, these studies highlight the power and versatility of RS and GIS in analysing and managing land resources.

LST is a valuable tool to monitor urban heat, assess climate change and its impact on human settlement, manage water bodies resources and its conservation. Recognizing the correlation between LULC and LST is crucial for addressing the challenges brought about by urban expansion (Siqi and Yuhong 2020). It provides essential insights into the complex interplay between human activities and the urban environment, offering potential solutions to urban growth-related issues. Given that LST is influenced by various land surface characteristics, its forecasting using LULC patterns becomes instrumental in the surveillance and alleviation of the detrimental consequences of urban expansion-induced temperature rise. When a robust correlation between LULC and LST is established, spectral indices can be effectively integrated into LULC models. This integration allows for the prediction and simulation of the potential ramifications of urban growth on the thermal conditions of a city's environment, aiding in informed decision-making and urban planning. However, there is a scarcity of studies that have examined the variations in LST and changes in LULC for the city of Raipur.

The primary aim of the research is to address specific scientific inquiries:

1. To measure the spatio-temporal transformations in LULC within Raipur city, focusing on decadal intervals (specifically for the years 2014, 2016, 2018, 2020, and 2022).

2. To assess the variations in LST during the summer seasons across the city, considering the temporal dimension.

3. To evaluate alterations in LST patterns in Raipur city attributable to the dynamics of LULC.

Study Area

Raipur, known as the 'Rice Bowl of India,' is a thriving city situated in the heart of Chhattisgarh state in central India. Raipur spans approximately between 21°23′N latitude to 21°50′N latitude and 81°38′E longitude to 82°00′E longitude, encompassing an area of around 226 km², as shown in Figure 1. One of Raipur's notable features is its significance as the state

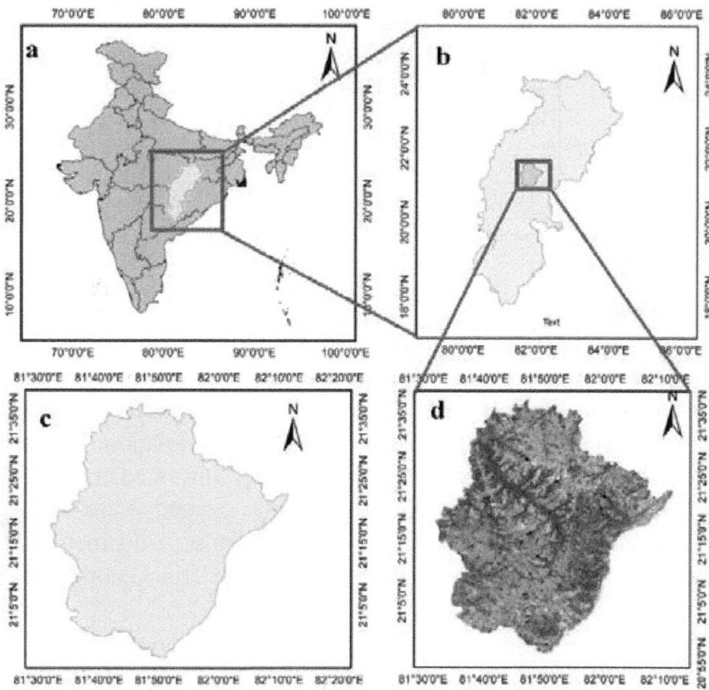

Figure 1. Study area of Raipur (a) Map of India, (b) Map of Chhattisgarh, (c) Administrative map of Raipur, (d) Satellite Image of Raipur.

capital of Chhattisgarh. The city plays a pivotal role in regional governance, education, commerce, and culture, fostering development and progress within the state. The city experiences a tropical wet and dry climate, characterised by an average annual rainfall of approximately 127.4 cm. Raipur experiences distinct seasons, with temperatures ranges from 5°C to 47°C during winter and summer. Relative humidity levels fluctuate between 20% and 80%, depending on the season.

Source of Data

Various satellite images were obtained from the USGS Data Centre which is shown in Table 1, LandSat 8 satellite series gives data, which has a resolution of 30 m. For all the satellite images path and row are mentioned for which it was used.

For research, USGS Earth Explorer was considered for downloading various data sets of Raipur.

- USGS Datasets (https://earthexplorer.usgs.gov/)

Table 1. Satellite image.

Year	Satellite	Resolution	Path/Row	Date
2014	LandSat 8	30 m	142/45	02-04-2014
2016	LandSat 8	30 m	142/45	23-04-2016
2018	LandSat 8	30 m	142/45	28-03-2018
2020	LandSat 8	30 m	142/45	02-04-2020
2022	LandSat 8	30 m	142/45	31-03-2021

Methodology

Extraction of LULC

Satellite imagery acquired from the USGS Earth Explorer and LandSat 8 OLI satellite series was employed for LULC determination. These images underwent a preprocessing phase to eliminate any unwanted errors, including cloud cover, and many bands were subsequently merged to create composite images. Depending on the specific analytical objectives, distinct bands were combined to generate composite images such as the True Color Composite or False Color Composite (FCC).

Subsequently, machine learning tools were applied to perform unsupervised classification. This involved the classification of pixels based on their unique Digital Number values, which were then associated with specific land types. In this particular study, LULC maps were categorized into four primary classes, namely water bodies, forest cover, urban area, and open land.

To evaluate the accuracy of this classification process, a verification test was conducted using Google Earth imagery. This test involved the assessment of 100 accuracy points by comparing them to the classified map. The results of the accuracy assessment relied on the allocation of training sample size and were expressed through an error matrix.

This comprehensive approach, encompassing satellite imagery, preprocessing, classification, accuracy assessment, and change detection analysis, provides valuable insights into the evolving trends in land use within the selected urban areas over the specified years.

Land Surface Temperature

In the context of satellite imagery, data is initially stored as Digital Numbers (DN), which serves as an indirect representation of the radiances emitted by ground objects. The calculation of LST is specifically derived from Band 10 and Band 11 in Landsat-8 images. A series of equations was used to get LST of Raipur.

Step 1: Calculating Top of Atmosphere denoted as (ToA) is done by Eq 1

$$ToA = M_B \ X \ BAND_{10} + A_B \tag{1}$$

where M_B is band specific of multiplicative rescaling factor, which in this case is 0.000334. A_B is Add band which is 0.10000. $BAND_{10}$ is the band no 10 which is obtained from satellite image.

Step 2: Calculating Brightness, which is given in Eq 2.

$$B_T = \left(\frac{K_2}{\ln\left(\frac{K_1}{L_\lambda}\right) + 1} \right) - 273.15 \tag{2}$$

where K_1 and K_2 are the Band Specific thermal conversion constant. $K_1 = 774.80$ and $K_2 = 1321.07$

Step 3: Calculation of NDVI which is given in Eq 3

$$NDVI = \frac{(Band\,5 - Band\,4)}{(Band\,5 + Band\,4)} \tag{3}$$

Step 4: Calculation of proportional vegetation P_V

$$P_v = \left(\frac{NDVI - NDVI_{min}}{NDVI_{max} - NDVI_{min}} \right)^2 \tag{4}$$

Step 5: Calculation of Emissivity (e)

$$e = 0.004 \ X \ P_v + 0.986 \tag{5}$$

Step 6: Calculation of LST is given by Eq 6

$$LST = \left(\frac{T_B}{1 + \left(\dfrac{0.00115 \ X \ T_B}{1.4388} \right) X \ln(e)} \right) \tag{6}$$

The above equation represents the equation for LST which is used in this study, with this equation spatial temperature was made for Raipur city for all years with time duration of 2 years.

Results and Discussion

Land use pattern

The following are LULC pattern for Raipur city which is depicted in below figure,

From Figure 2, it is evident that machine learning techniques were employed to classify the Raipur city using satellite images. In this case, unsupervised classification was utilized, resulting in a commendable level of accuracy. The land uses were categorized into four distinct classes, a color matrix was generated, offering valuable insights: blue represents water bodies, green signifies forests and vegetation cover, red corresponds to urban built-up areas, and yellow denotes open land primarily used for agriculture.

Upon closer analysis, it becomes apparent that water bodies have remained relatively stable over time, constituting less than 3 percent of the entire region, which is cause for concern.

On the other hand, urban areas have consistently expanded at a steady pace. However, there has been a decline in forested areas from 2014 to 2022. The reduction in open land can be attributed to urbanization, as these areas are progressively transformed to support various developmental activities.

The above Figure 3 illustrates, Over the years spanning from 2014 to 2022, the land use dynamics within the studied area have exhibited some noteworthy

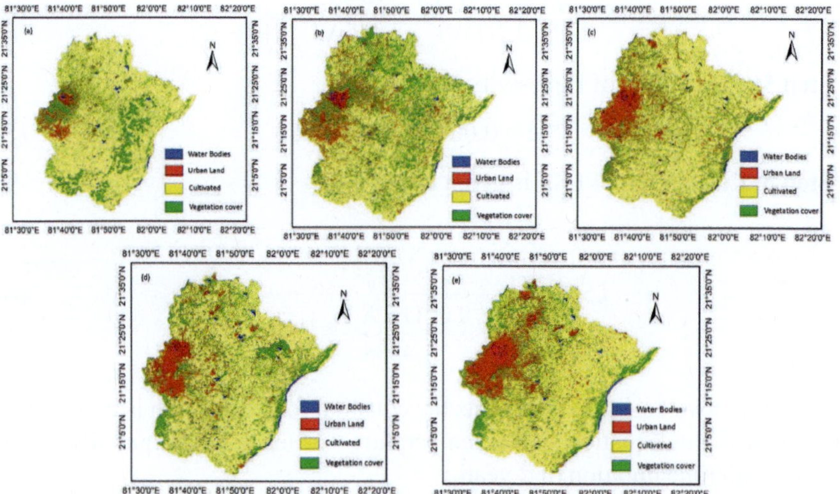

Figure 2. Land use pattern for years (a) 2014, (b) 2016, (c) 2018, (d) 2020, (e) 2022 of Raipur.

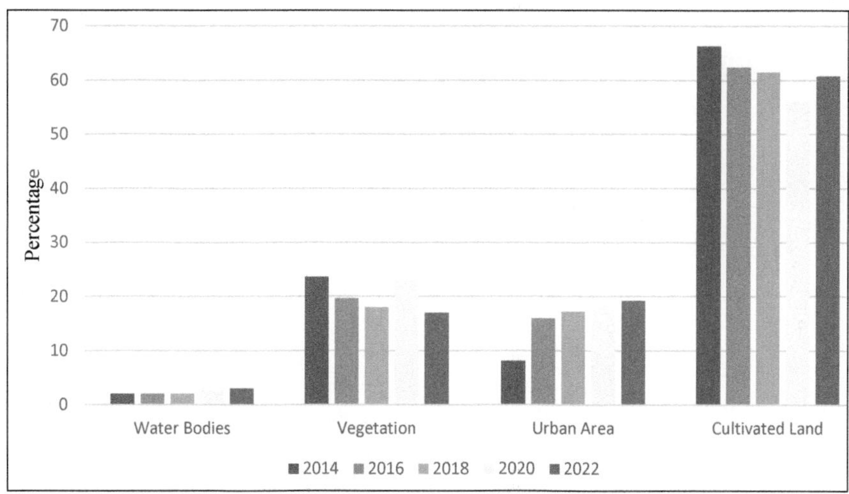

Figure 3. Change in land use pattern of Raipur for different year.

trends. The data indicates that water bodies have shown a gradual increase in coverage, starting at 1.97% in 2014 and steadily rising to 3% by 2022. This may signify certain environmental changes or alterations in water resources. Conversely, the area covered by vegetation experienced a fluctuating pattern, with a notable decrease from 23.6% in 2014 to 17% in 2022. This change might be attributed to factors such as deforestation, urbanization, or natural variations in vegetation. Urban areas displayed consistent growth, increasing from 8.1% in 2014 to 19.2% in 2022, indicating ongoing urban development and expansion within the region. The percentage of open land, which is utilized for a variety of uses, decreased from 66.3% in 2014 to 60.8% in 2022. This decrease may have been brought about by the conversion of open land for urban and development projects.

Spatial distribution of land surface temperature

The Figure 4, provides a visual representation of LST. This image depicts the spatial distribution of temperatures in degrees Celsius (°C). The temperature data has been categorized into five distinct ranges, each represented by a specific color code. Notably, the darker shades of green indicate the lowest temperatures, while the darker red hues signify the highest temperatures within this classification.

A comprehensive analysis of the figure reveals some intriguing patterns. It is evident that areas with a substantial presence of vegetation cover tend

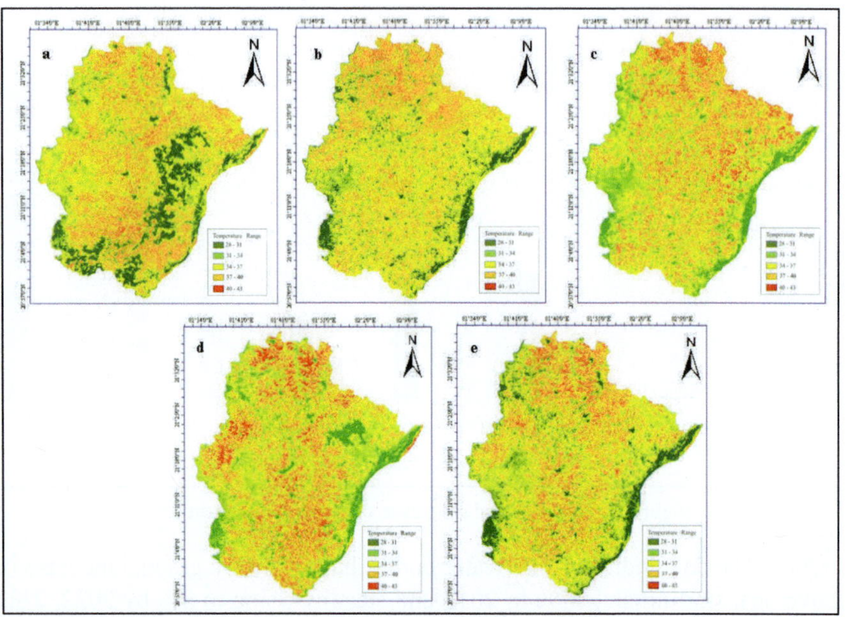

Figure 4. The figure represents Land Surface Temperature of Raipur, for years (a) 2014, (b) 2016, (c) 2018, (d) 2020, (e) 2022.

to exhibit comparatively lower temperatures. Conversely, open lands or bare areas register higher temperatures. This correlation underscores the moderating effect of vegetation on local temperatures, as it provides shade and evaporative cooling, thereby lowering the LST. On the other hand, open, exposed areas absorb more solar radiation, resulting in elevated land surface temperatures. This information can be used for environmental management, sustainable development, management of resourced and many more.

Conclusion

The present study LST and land use patterns of Raipur City is determined, by using Landsat data sets of four different time durations 2014, 2016, 2018, 2020 and 2022.

Data suggests dynamic changes in land use over this period, with increasing water bodies and urban areas, a decrease in vegetation, and a decline in open land. These trends may reflect the evolving socio-economic and environmental landscape of the studied area. Further analysis and investigation are required to understand the underlying causes and potential implications of these shifts in land use.

The study provides clear insight in exploration of the dynamics in LST across various LULC categories in the study area, considering a decadal perspective. Notably, the results of this analysis underscore an important and consistent increment in urban areas over the past years, surpassing the growth observed in other land cover categories. This expansion in built-up regions can be attributed to the conversion of other areas into urban developments.

Furthermore, the analysis of LST variations of different land-use classes reveals that barren land, agricultural areas, open areas and urban regions consistently exhibit higher temperatures compared to the area of vegetation cover categories.

The insights derived from this research contribute to a solid scientific foundation for the sustainable and efficient management of land resources. This study is focused to serve as a valuable resource for urban developers and policymakers, empowering them to get a clear insight of impact of urbanization and make informed decisions aimed at enhancing the sustainability of cities.

References

Arowolo, A. O. and Deng, X. 2018. Land use/land cover change and statistical modelling of cultivated land change drivers in Nigeria. Regional Environmental Change 18: 247–259.

Fu, P. and Weng, Q. 2016. A time series analysis of urbanization induced land use and land cover change and its impact on land surface temperature with Landsat imagery. Remote Sensing of Environment 175: 205–214.

Guha, S., Govil, H., Taloor, A. K., Gill, N. and Dey, A. 2022. Land surface temperature and spectral indices: A seasonal study of Raipur City. Geodesy and Geodynamics 13(1): 72–82.

Haldar, S., Mandal, S., Bhattacharya, S. and Paul, S. 2023. Dynamicity of land use/land cover (LULC): An analysis from peri-urban and rural neighbourhoods of Durgapur Municipal Corporation (DMC) in India. Regional Sustainability 4(2): 150–172.

Liu, J., Wang, Z., Duan, Y., Li, X., Zhang, M., Liu, H. et al. 2023. Effects of land use patterns on the interannual variations of carbon sinks of terrestrial ecosystems in China. Ecological Indicators 146: 109914.

Martins, I. S., Proença, V. and Pereira, H. M. 2014. The unusual suspect: Land use is a key predictor of biodiversity patterns in the Iberian Peninsula. Acta Oecologica 61: 41–50.

Mu, L., Wang. L., Wang, Y., Chen, X. and Han, W. 2019. Urban land use and land cover change prediction via self-adaptive cellular based deep learning with multisourced data. IEEE Journal of Selected Topics in Applied Earth Observations and Remote Sensing 12(12): 5233–5247.

Prishchepov, A. V., Radeloff, V. C., Dubinin, M. and Alcantara, C. 2012. The effect of Landsat ETM/ETM+ image acquisition dates on the detection of agricultural land abandonment in Eastern Europe. Remote Sensing of Environment 126: 195–209.

Roy, B. and Bari, E. 2022. Examining the relationship between land surface temperature and landscape features using spectral indices with Google Earth Engine. Heliyon 8(9).

Seyam, M. M. H., Haque, M. R. and Rahman. M. M. 2023. Identifying the land use land cover (LULC) changes using remote sensing and GIS approach: A case study at Bhaluka in Mymensingh, Bangladesh. Case Studies in Chemical and Environmental Engineering 7: 100293.

Siqi, J. and Yuhong, W. 2020. Effects of land use and land cover pattern on urban temperature variations: A case study in Hong Kong. Urban Climate 34: 100693.

Wang, J., Wei, H., Cheng, K., Ochir, A., Davaasuren, D., Li, P. et al. 2020. Spatio-temporal pattern of land degradation from 1990 to 2015 in Mongolia. Environmental Development 34: 100497.

Wang, L., Wang, S., Zhou, Y., Zhu, J., Zhang, J., Hou, Y. et al. 2020. Landscape pattern variation, protection measures, and land use/land cover changes in drinking water source protection areas: A case study in Danjiangkou Reservoir, China. Global Ecology and Conservation 21: e00827.
Wu, K. Y. and Zhang, H. 2012. Land use dynamics, built-up land expansion patterns, and driving forces analysis of the fast-growing Hangzhou metropolitan area, eastern China (1978–2008). Applied Geography 34: 137–145.

8

Exploring the Integration of Machine Learning for Environmental Pollution and Flood Risk Assessment

A Comprehensive Review

Padam Jee Omar,[1,]* *Shashank Singh,*[2]
Purushottam Kumar Mahato[2] *and Ghazaala Yasmin*[3]

Introduction

Flood prediction and management are vital due to their global impact that cause humanitarian crises, economic losses, and environmental damage (Wang et al. 2015). This necessitates effective strategies and technologies for mitigation.

Background and significance

Worldwide, floods are a concern due to both natural and human-caused sources (Tehrany et al. 2015). The implications can be humanitarian, economic, and

[1] Department of Civil Engineering, Babasaheb Bhimrao Ambedkar University (A central University), Lucknow-226025.
[2] Floodkon Consultants LLP, Noida-201304.
[3] Dept. of Computer Science and Engineering/Information Technology, Jaypee Institute of Information Technology.
* Corresponding author: drpadamomar@gmail.com

environmental, necessitating community involvement, adapting to climate change, and building robust infrastructure (Rashid 2018).

Overview of machine learning algorithms in water resources engineering

Algorithms for machine learning are essential for improving flood control and prediction (Cannas et al. 2005). They make it possible to precisely analyse intricate hydrological data, which supports early warning systems and water resources engineering decision-making (Merz et al. 2010, Omar and Kumar 2021).

Research Objectives and Methodology

The objective of the study is to investigate and apply machine learning methods to enhance flood forecasting. The methodology for more precise and timely flood management techniques entails the study of historical data, creating forecasting models, and evaluating their effectiveness.

Data Collection and Preparation

Collection of historical weather data, river levels, and relevant variables

Weather data: Acquiring data on temperature, humidity, wind patterns, and precipitation is necessary in order to compile historical weather information. Satellites, weather archives, and meteorological stations all offer useful datasets for study (Abbot and Marohasy 2014).

River levels: Predicting floods depends on keeping an eye on river levels. Information on water levels over time can be gathered via hydrological models, remote sensing technology, and river gauge stations (Kasiviswanathan et al. 2016).

Important factors: Flood dynamics are influenced by a number of factors, including topography, soil type, and land use, in addition to weather and river levels. Gathering extensive datasets guarantees a full comprehension of the variables impacting floods (Gupta et al. 2022b).

Data Preprocessing and Quality Control Techniques

Cleaning and imputation: It is critical to make a note of incomplete or erroneous data. The integrity of the dataset is maintained using methods like imputation and the elimination of incomplete records.

Normalisation and scaling: Creating uniform scales and units for various variables makes integrating and comparing disparate datasets easier.

Outlier identification and management: Preserving data accuracy requires the identification and control of outliers. Statistical techniques can be used to modify or eliminate extreme values. Aggregating data both geographically (from grid cells to watershed level) and temporally (from hourly to daily, for example) can improve computing efficiency and model performance (Gupta et al. 2022a).

Quality assurance inspections: The accuracy, consistency, and logical coherence of data are checked to guarantee the dataset's dependability.

Feature engineering: Increasing the predictive capacity of models may be achieved by adding new features or altering current ones. For example, using raw data to derive variables like river flow rate or rainfall intensity. Accurate and dependable flood prediction models are based on well-executed data preparation and collecting. The study can provide insights into the intricate relationships impacting flood occurrences by combining a variety of well-pre-processed information, facilitating more informed decision-making in the management of water resources.

Machine Learning Algorithms for Flood Prediction

Overview of commonly used machine learning algorithms

Commonly used machine learning algorithms include Linear Regression for continuous prediction, Logistic Regression for binary classification, Decision Trees for multidimensional classification and regression, Random Forest for improved accuracy through ensemble learning, Support Vector Machines for effective high-dimensional classification, K-Nearest Neighbors for simple yet effective classification and regression. Figure 1 shows the basic flow diagram for choosing the machine learning model.

Random forests: An ensemble learning technique that combines several decision trees to provide reliable forecasts (Ni et al. 2020).

Support Vector Machines (SVM): Determine the best hyperplane in a high-dimensional space to classify data points (Lin et al. 2006, Sankaranarayanan et al. 2020).

Neural networks: Emulate the structure of the human brain to identify intricate patterns in data (Gaur et al. 2023).

Gradient boosting machines: Constructing a sequence of weakened models, each of which fixes the flaws in the preceding one (Dubossarsky et al. 2016).

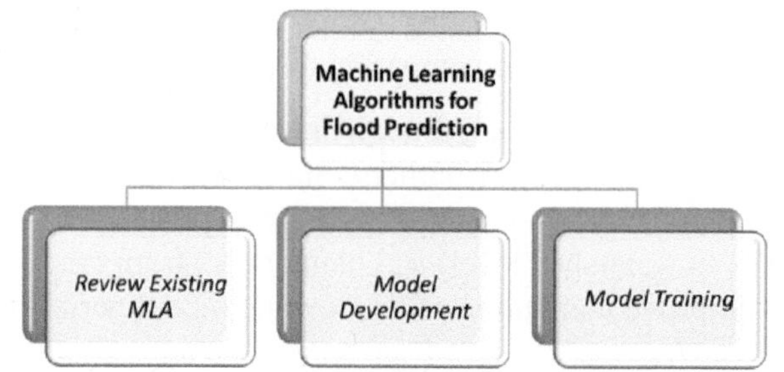

Figure 1. Basic flow diagram for building the machine learning (ML) model.

K-nearest neighbours (KNN): This algorithm groups data points in the feature space according to the predominant class of their neighbours (Modaresi et al. 2018).

XGBoost: A potent gradient boosting algorithm with a solid track record of efficiency and speed (Ma et al. 2021, Kaur et al. 2021).

Decision trees: Creating a decision-making structure analogous to a tree by segmenting data into subsets according to feature values (Khosravi et al. 2018).

Recurrent Neural Networks (RNN): Because they can capture temporal relationships, RNNs are especially helpful for sequential data, such time series (Chang et al. 2014).

Feature Selection and Model Development

Analysing the significance of each input variable aids in the selection of the most important characteristics for precise prediction.

Dimensionality reduction: The number of features may be decreased without sacrificing important information by using methods like Principal Component Analysis (PCA) or t-Distributed Stochastic Neighbour Embedding (t-SNE).

Model selection: Selecting the best method according to the needs of the problem and the properties of the dataset. Optimising algorithm parameters to enhance model performance and generalisation is known as hyperparameter tuning.

Ensemble methods: Predictive accuracy and resilience can be improved by combining numerous models (such as bagging or stacking).

Training and Validation of Flood Prediction Models

Data splitting: To guarantee that the model is trained on one subset and tested on another, evaluating its generalisation capacity, the dataset is divided into training and validation sets.

Cross-validation: Using methods such as k-fold cross-validation yields a more reliable assessment of the model's operation.

Metrics for evaluation: Assessing the model's accuracy, precision, recall, and other pertinent parameters using suitable metrics, such as Mean Squared Error (MSE) or Receiver Operating Characteristic (ROC) curves.

Preventing overfitting: In neural networks, regularisation strategies and dropout layers aid in preventing overfitting, ensuring that the model performs effectively when applied to new data (Rajurkar et al. 2004).

Constant monitoring and updating: Adding fresh data to models on a regular basis keeps them current and flexible in response to shifting circumstances.

Early Warning Systems

Integration of predictive models into early warning systems

Model integration: Predictive power is increased in early warning systems by integrating machine learning models. These models predict possible flood occurrences based on both historical and current data.

Decision Support Systems (DSS): Making decisions based on model outputs may be done with knowledge when predictive models are integrated with DSS. This guarantees that alerts are customised to particular circumstances and hazards.

Real-time Data Acquisition and Processing

Sensor networks: Real-time data collection is made possible by utilising a network of sensors, such as weather stations, river gauges, and other environmental monitoring equipment. The utilisation of satellite images and remote sensing technology facilitates the continual monitoring of evolving environmental conditions by offering a wider viewpoint (Rezaeian-Zadeh et al. 2013).

Internet of Things (IoT) devices: Smart sensors and other IoT devices add to a dynamic, networked data environment that speeds up data collecting.

Timely Alerts and Notifications to Relevant Stakeholders

Automated alert systems: Putting automated alert systems in place guarantees that warnings are sent out on time. Predefined thresholds obtained from prediction models may be used to initiate these notifications as shown in Figure 2.

Communication channels: They ensure that warnings are received by a large audience by using a variety of communication channels, such as social media, mobile applications, SMS, and sirens.

Stakeholder engagement: A coordinated response is ensured by including government agencies, emergency responders, and local communities in the early warning system. Community exercises and training initiatives improve readiness.

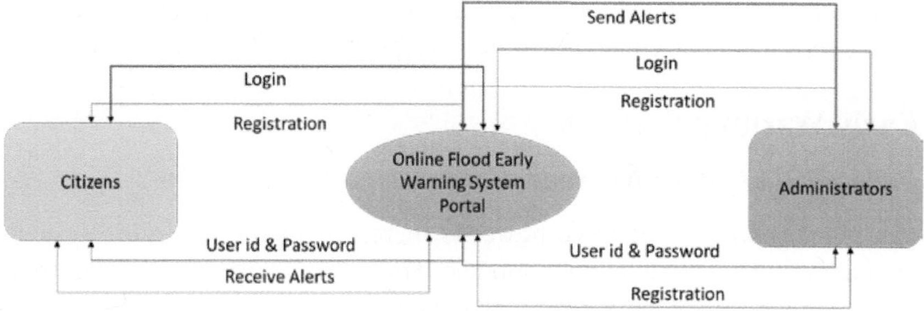

Figure 2. Key steps for online flood early warning system.

Continuous Monitoring and Adaptation

Feedback loops: Continuous improvement is made possible by creating feedback loops between the prediction models and the early warning system. Refinement of the model performance can be achieved by measuring how well the system predicts and mitigates flood disasters.

Adaptive thresholds: This feature keeps the system sensitive to changing hazards by adjusting warning thresholds in response to shifting environmental variables and the prediction models' performance (Omar et al. 2017).

Planning scenarios: Taking into account different flood situations and their possible effects, aids in fine-tuning warning levels and reaction tactics.

Flood Risk Assessment

Utilising machine learning models for flood risk assessment

Predictive modelling: To forecast probable flood episodes, machine learning algorithms use both historical and current data (Kisi et al. 2012). To assess the possibility of flooding, these models take into account a variety of parameters, including rainfall, river levels, and terrain features (Araghinejad et al. 2011).

Measuring risk variables: Machine learning algorithms are able to measure the effect of many risk variables on floods, offering a comprehensive picture of the intricate relationships that influence flood risk.

Scenario analysis: These models may evaluate the possible effects of varying flood magnitudes through the use of simulations and scenario analyses, which helps with risk assessment.

Integration with Geographic Information Systems (GIS)

Geographical analysis: To visualise and understand geographical interactions, GIS incorporates a variety of variables, such as topography, land use, and infrastructure. Assessments of flood risk are now more accurate because to this integration (Omar et al. 2021a).

Mapping vulnerabilities: A full picture of the possible effects of flooding may be obtained by using GIS to map and analyze spatially sensitive regions, vital infrastructure, and population density.

Decision support: Using GIS, stakeholders may visualize and understand complicated geographical data to make well-informed decisions on flood risk management (Liang et al. 2018).

Identification of High-Risk Areas and Vulnerable Infrastructure

Risk zoning: By integrating machine learning models with geographic information systems (GIS), risk zones may be established according to the probability and possible severity of floods. Development codes and land-use planning are guided by these zones.

Infrastructure vulnerability assessment: By examining how vulnerable vital infrastructure—like hospitals, power plants, and transportation networks—is, it becomes easier to allocate funds for defence and resilience strategies.

Community exposure: Determining locations of high population density or socioeconomic susceptibility to floods offers valuable information about the possible effects on human life, which can inform evacuation strategies and resource allocation (Tehrany et al. 2015).

Optimization Strategies for Flood Management

Machine learning-based optimisation techniques

Optimal resource allocation: By examining historical data and forecasting the areas where interventions are most needed, machine learning algorithms can optimise the allocation of resources for flood management. This entails effectively deploying barriers, emergency services, and relief operations.

Real-time decision support: By incorporating machine learning into decision support systems, flood management techniques may be analysed and adjusted in real-time. Algorithms can adjust to shifting circumstances, increasing the overall efficacy of response initiatives (Jee et al. 2019).

Predictive maintenance: By identifying when equipment is likely to break, machine learning models can optimise flood control infrastructure maintenance. By taking this preventive measure, downtime is reduced and flood management system dependability is guaranteed.

Resource allocation and emergency response planning

Dynamic resource allocation: Based on real-time data and predictive analytics, optimisation models may distribute resources—such as staff, supplies, and equipment—dynamically. This guarantees a more efficient and flexible reaction to changing flood situations.

Scenario-based planning: Strategic resource planning is made possible by creating scenarios based on various flood durations and magnitudes. Resource allocation might be based on the identification of the most crucial regions using optimisation models.

Interagency collaboration: When it comes to emergency response, coordination between the different agencies is essential. Optimisation techniques can help with cooperative planning, guaranteeing effective resource usage and a coordinated response.

Infrastructure Design and Floodplain Management

Optimal infrastructure design: By examining previous flood data and forecasting future hazards, machine learning algorithms may help with the design of infrastructure that is resistant to flooding. This entails making levees, dams, and other defensive structures as tall and functional as possible.

Floodplain zoning: By determining appropriate land uses and zoning laws, GIS and machine learning may be utilised to optimise floodplain management. By doing so, unsuitable growth in high-risk locations is avoided, and sustainable land use techniques are encouraged (Omar et al. 2021b).

Natural infrastructure solutions: Natural infrastructure solutions, such wetlands and green areas, may be included into flood management plans as part of optimisation techniques. The success of these measures in lowering the danger of flooding may be evaluated via machine learning.

Case Studies and Applications

Successful implementation

IBM's GRAF predicts floods accurately by utilising machine learning. It forecasts extreme weather occurrences by analysing massive amounts of atmospheric data. With the use of real-time data and sophisticated modelling, GRAF generates accurate and timely forecasts.

Real-World Examples

In order to accurately anticipate floods, Google's programmes in India employs machine learning to assess geography, weather, and river levels. It permits prompt notifications, supporting proactive actions by authorities.

Applications

Applications of machine learning (ML) in soil and land flood risk modelling provide creative ways to anticipate, assess, and lessen the effects of floods on these vital environmental components (Choubin et al. 2018).

Here are a few particular uses:

Terrain analysis and land cover classification: In order to categorise different types of land cover and the features of the terrain, machine learning algorithms can examine elevation data and satellite images. Understanding how various terrain types react to floods and identifying locations that are more vulnerable depending on their features require knowledge of this information.

Soil permeability and saturation prediction: Soil permeability and saturation levels may be predicted by machine learning models by using groundwater levels, soil composition data, and historical weather records. Comprehending the pace at which soil is saturated and loses its capacity to absorb water is essential for evaluating the likelihood of runoff and severe flooding.

Flood inundation mapping: Spatial data may be processed and analysed by ML algorithms to produce intricate maps of flood inundation (Elsafi 2014). These maps aid in disaster response and land-use planning by giving a visual depiction of the regions that are most likely to experience floods (Chang et al. 2014).

Risk assessment and vulnerability analysis: To determine flood risk and susceptibility, machine learning models use a number of variables, including geography, past flood data, and soil properties. This makes it possible for decision-makers to more efficiently allocate resources, put preventive measures into place, and prioritise areas for action (Tripathi and Pandey 2022).

Real-time monitoring and early warning systems: Real-time weather conditions, river levels, and soil moisture may all be continually monitored by ML-driven systems. Authorities and communities may be informed about impending flood dangers using early warning systems driven by machine learning, enabling prompt evacuation and preventative measures.

Post-flood impact assessment: Once an incident has occurred, machine learning (ML) may help evaluate how floods affect the land and soil. This includes assessing land cover changes, sediment deposition, soil erosion, and potential long-term impacts on ecosystems and agriculture.

Dynamic decision support systems: Flood risk models may be constantly adjusted by ML algorithms in response to shifting environmental factors. This data may be used by decision support systems to suggest in-the-moment flood risk management strategies, such as modifying dam water release rates or erecting temporary barriers.

Application of ML in water pollution: Water pollution issues are addressed by machine learning (ML) applications in soil and land flood risk modelling. Algorithms using machine learning (ML) can track water quality measures, locate pollution sources, forecast the spread of contaminants during floods, and enhance cleanup tactics (Cao et al. 2020). ML models provide ecological impact assessment, adaptive water quality management, and early warning systems for water pollution (Taoufik et al. 2022). By taking into account water pollution, environmental risk management becomes more comprehensive and ecosystems and communities are more resilient to the multiple problems that floods provide.

Challenges and Limitations

Data availability and quality

Challenge: Limited and inconsistent data: Getting enough high-quality data is a recurring problem in flood prediction. Real-time data availability can fluctuate, and historical data used to train models may not be fully available. There could be gaps in the datasets due to inadequate infrastructure for data collection in some areas.

Limitation: biases and inaccuracies: Predictions can be erroneous as a result of incomplete or biased datasets. The accuracy of flood projections may be impacted if models trained on insufficient data are unable to fully represent the complexity of local circumstances.

Uncertainty and Model Accuracy

Challenge: inherent uncertainty in weather patterns: Because weather systems are inherently variable, developing completely accurate flood prediction models is difficult. Although machine learning has the potential to improve accuracy, uncertainty cannot be avoided due to the dynamic and complicated nature of atmospheric circumstances.

Limitation: Over-reliance on historical patterns: Models that just take into account past data could find it difficult to adjust to novel situations. The models' capacity to reliably forecast extreme weather occurrences may be hampered by the introduction of new patterns brought about by climate change and other dynamic variables.

Integration with Existing Flood Management Systems

Challenge: seamless integration with traditional systems: It can be difficult to integrate machine learning models with current flood control methods. Challenges might arise from the requirement for real-time data interchange between various system components and compatibility difficulties.

Limitation: resistance to technological adoption: It is possible that certain current systems will not accept future technology. A major obstacle may be getting stakeholders to trust and feel at ease with the machine learning models in the face of institutional inertia.

Conclusion

In this study, effective machine learning applications for flood control and prediction, have been examined showcasing projects like Google's India programme and IBM's GRAF system. These systems use real-time data analysis and sophisticated modelling to produce flood forecasts that are more precise and timelier. We also spoke about the difficulties, such as the lack of consistency in weather patterns, the difficulty of obtaining high-quality data, and the difficulty of integrating machine learning with the current flood control systems.

Potential Impact of Machine Learning in Flood Prediction and Management

Machine learning has a significant potential influence on flood control and prediction. Machine learning makes it possible to anticipate floods more precisely and promptly by utilising the capabilities of predictive modelling and data analysis. Proactive actions like early warning systems, effective resource allocation, and enhanced catastrophe preparedness are therefore made possible. With this technique, the effects of floods on infrastructure, communities, and the environment might be greatly mitigated.

Recommendations for Further Research and Implementation

Enhance data infrastructure: Solving data problems is essential. Enhancing data infrastructure and guaranteeing the availability of extensive and superior datasets for machine learning model training should be the main goals of future research.

Continuous model refinement: With weather patterns being dynamic and including inherent uncertainties, future research should focus on improving machine learning model accuracy. The goal is to continuously develop the model and adapt it to changing environmental conditions.

Collaborative approaches: Efforts should be made to promote cooperation among meteorologists, government agencies, local communities, and technological specialists. Diverse stakeholders working together guarantees that machine learning models are more palatable and in line with local conditions.

To sum up, machine learning has enormous potential to transform flood control and prediction. Even if there are obstacles, it is a desirable field for ongoing study and implementation efforts because of the potential advantages in terms of improved forecast accuracy and proactive mitigation actions. The use of machine learning in flood management may greatly improve community resilience and lessen the effects of floods globally by tackling the issues that have been discovered and implementing the suggested solutions.

References

Abbot, J. and Marohasy, J. 2014. Input selection and optimisation for monthly rainfall forecasting in Queensland, Australia, using artificial neural networks. Atmospheric Research 138: 166–178.

Araghinejad, S., Azmi, M. and Kholghi, M. 2011. Application of artificial neural network ensembles in probabilistic hydrological forecasting. Journal of Hydrology 407(1-4): 94–104.

Cannas, B., Fanni, A., Sias, G., Tronci, S. and Zedda, M. K. 2005. River flow forecasting using neural networks and wavelet analysis. In Geophys. Res. Abstr 7: 08651.

Cao, X., Liu, Y., Wang, J., Liu, C. and Duan, Q. 2020. Prediction of dissolved oxygen in pond culture water based on K-means clustering and gated recurrent unit neural network. Aquacultural Engineering 91: 102122.

Chang, F. J., Chen, P. A., Lu, Y. R., Huang, E. and Chang, K. Y. 2014. Real-time multi-step-ahead water level forecasting by recurrent neural networks for urban flood control. Journal of Hydrology 517: 836–846.

Chang, L. C., Shen, H. Y. and Chang, F. J. 2014. Regional flood inundation nowcast using hybrid SOM and dynamic neural networks. Journal of Hydrology 519: 476–489.

Choubin, B., Darabi, H., Rahmati, O., Sajedi-Hosseini, F. and Kløve, B. 2018. River suspended sediment modelling using the CART model: A comparative study of machine learning techniques. Science of the Total Environment 615: 272–281.

Dubossarsky, E., Friedman, J. H., Ormerod, J. T. and Wand, M. P. 2016. Wavelet-based gradient boosting. Statistics and Computing 26: 93–105.

Elsafi, S. H. 2014. Artificial neural networks (ANNs) for flood forecasting at Dongola Station in the River Nile, Sudan. Alexandria Engineering Journal 53(3): 655–662.

Gaur, S., Omar, P. J. and Eslamian, S. 2023. Advantage of grid-free analytic element method for identification of locations and pumping rates of wells. pp. 1–10. *In*: Eslamian, S. and Faezeh Eslamian (eds.). Handbook of HydroInformatics: Volume III: Water Data Management Best Practices. Elsevier, doi.org/10.1016/B978-0-12-821962-1.00003-9.

Gupta, N. Mahato, P. K., Patel, J., Omar, P. J. and Tripathi, R. P. 2022a. Understanding trend and its variability of rainfall and temperature over Patna (Bihar). pp. 533–543. *In*: Zakwan, M., Wahid, A., Niazkar, M. and Chatterjee, U. (eds.). Current Directions in Water Scarcity Research, Vol. 7, Elsevier, doi.org/10.1016/B978-0-323-91910-4.00030-3.a

Gupta, N., Patel, J., Gond, S., Tripathi, R. P., Omar, P. J. and Dikshit, P. K. S. 2022b. Projecting future maximum temperature changes in river ganges basin using observations and statistical downscaling model (SDSM). pp. 561–585. *In*: Pandey, M., Azamathulla, H. and Pu, J. H. (eds.). River Dynamics and Flood Hazards: Disaster Resilience and Green Growth. Springer Nature Singapore. doi.org/10.1007/978-981-19-7100-6_31.

Jee, O. P., Bihari, D. S. and Kumar, D. P. 2019. Temporal variability study in rainfall and temperature over Varanasi and adjoining areas. Disaster Advances 12(1): 1–7.

Kasiviswanathan, K. S., He, J., Sudheer, K. P. and Tay, J. H. 2016. Potential application of wavelet neural network ensemble to forecast streamflow for flood management. Journal of Hydrology 536: 161–173.

Kaur, R., Schaye, C., Thompson, K., Yee, D. C., Zilz, R., Sreenivas, R. S. et al. 2021. Machine learning and price-based load scheduling for an optimal IoT control in the smart and frugal home. Energy and AI 3: 100042.

Khosravi, K., Pham, B. T., Chapi, K., Shirzadi, A., Shahabi, H., Revhaug, I. et al. 2018. A comparative assessment of decision trees algorithms for flash flood susceptibility modeling at Haraz watershed, northern Iran. Science of the Total Environment 627: 744–755.

Kisi, O., Nia, A. M., Gosheh, M. G., Tajabadi, M. R. J. and Ahmadi, A. 2012. Intermittent streamflow forecasting by using several data driven techniques. Water Resources Management 26: 457–474.

Liang, Z., Li, Y., Hu, Y., Li, B. and Wang, J. 2018. A data-driven SVR model for long-term runoff prediction and uncertainty analysis based on the Bayesian framework. Theoretical and Applied Climatology 133: 137–149.

Lin, J. Y., Cheng, C. T. and Chau, K. W. 2006. Using support vector machines for long-term discharge prediction. Hydrological Sciences Journal 51(4): 599–612.

Ma, M., Zhao, G., He, B., Li, Q., Dong, H., Wang, S. et al. 2021. XGBoost-based method for flash flood risk assessment. Journal of Hydrology 598: 126382.

Merz, B., Hall, J., Disse, M. and Schumann, A. 2010. Fluvial flood risk management in a changing world. Natural Hazards and Earth System Sciences 10(3): 509–527.

Modaresi, F., Araghinejad, S. and Ebrahimi, K. 2018. A comparative assessment of artificial neural network, generalized regression neural network, least-square support vector regression, and K-nearest neighbor regression for monthly streamflow forecasting in linear and nonlinear conditions. Water Resources Management 32(1): 243–258.

Ni, L., Wang, D., Wu, J., Wang, Y., Tao, Y., Zhang, J. et al. 2020. Streamflow forecasting using extreme gradient boosting model coupled with Gaussian mixture model. Journal of Hydrology 586(124): 901.

Omar, P. J., Gupta, N., Tripathi, R. P. and Shekhar, S. 2017. A study of change in agricultural and forest land in Gwalior city using satellite imagery. SAMRIDDHI: A Journal of Physical Sciences, Engineering and Technology 9(02): 109–112.

Omar, P. J. and Kumar, V. 2021. Assessment of damage for dam break incident in Lao PDR using SAR data. Int. J. Hydrol. Sci. Technol. https://doi.org/10.1504/IJHST.2021.10040874.

Omar, P. J., Gaur, S., Dwivedi, S. B. and Dikshit, P. K. S. 2021a. Development and application of the integrated GIS-MODFLOW model. pp. 305–314. *In*: Gupta, P. K. and Bharagava, R. N. (eds.). Fate and Transport of Subsurface Pollutants. Microorganisms for Sustainability, vol 24, Springer Nature, Singapore. doi.org/10.1007/978-981-15-6564-9_16.

Omar, P. J., Shivhare, N., Dwivedi, S. B., Gaur, S. and Dikshit, P. K. S. 2021b. Study of methods available for groundwater and surfacewater interaction: a Case Study on Varanasi, India. pp. 67–83. *In*: Chauhan, M. S. and Ojha, C. S. P. (eds.). The Ganga River Basin: A Hydrometeorological Approach. Society of Earth Scientists Series. Springer Nature, Switzerland. doi.org/10.1007/978-3-030-60869-9_5.

Rajurkar, M. P., Kothyari, U. C. and Chaube, U. C. 2004. Modeling of the daily rainfall-runoff relationship with artificial neural network. Journal of Hydrology 285(1-4): 96–113.

Rashid, K. 2018. Flood hazard mapping for the humanitarian sector: an opinion piece on needs and views. Global Flood Hazard: Applications in Modeling, Mapping, and Forecasting 115–130.

Rezaeian-Zadeh, M., Tabari, H. and Abghari, H. 2013. Prediction of monthly discharge volume by different artificial neural network algorithms in semi-arid regions. Arabian Journal of Geosciences 6: 2529–2537.

Sankaranarayanan, S., Prabhakar, M., Satish, S., Jain, P., Ramprasad, A. and Krishnan, A. 2020. Flood prediction based on weather parameters using deep learning. Journal of Water and Climate Change 11(4): 1766–1783.

Taoufik, N., Boumya, W., Achak, M., Chennouk, H., Dewil, R. and Barka, N. 2022. The state of art on the prediction of efficiency and modeling of the processes of pollutants removal based on machine learning. Science of The Total Environment 807: 150554.

Tehrany, M. S., Pradhan, B. and Jebur, M. N. 2015. Flood susceptibility analysis and its verification using a novel ensemble support vector machine and frequency ratio method. Stochastic Environmental Research and Risk Assessment 29: 1149–1165.

Tripathi, R. P. and Pandey, K. K. 2022. Numerical investigation of flow field around T-shaped spur dyke in a reverse-meandering channel. Water Supply 22(1): 574–588.

Wang, Z., Lai, C., Chen, X., Yang, B., Zhao, S. and Bai, X. 2015. Flood hazard risk assessment model based on random forest. Journal of Hydrology 527: 1130–1141.

9

Geo-AI for Urban Planning

Kumari Pritee[1,*] *and R. D. Garg*[2]

Introduction

Artificial geospatial intelligence (GeoAI) has received a lot of attention recently. GeoAI is an emerging scientific field that combines innovations in spatial science and AI techniques: such as ML and DL, data mining, and high-performance computing. Its goal is to help organize the way we think about and approach spatial big data processing and analysis. GeoAI, is the use of AI techniques such as ML and DL to interrogate geographic data and imagery to generate knowledge (Ferneda et al. 2016). GeoAI is growing in importance and promise due to the increasing availability of geographic data, AI developments, and vast computing power. This idea is positioned within the larger framework of AI intra connections (Kour et al. 2023) a subfield of AI that applies machine learning to extract information from geographic data.

GeoAI systems now play a significant role in enhancing traditional AI technologies and creating innovative methods to handle certain issues given by geospatial data, which is regarded as geo-referenced data containing position markers, are vast, intricate, diverse, and growing.

As we can see, geospatial data are used in a wide range of scientific and practical fields, including public health, safety, smart cities, transportation, business, resilience to natural catastrophes, and climate change. The ultimate objective is to enhance the standard of living for the world's expanding urban population by addressing these concerns. These multidisciplinary

---

[1] Assistant Professor, Information System Management, IIM Sambalpur.

[2] Professor, Geomatics Engineering, IIT Roorkee, Roorkee, India.

* Corresponding author: preetik@iimsambalpur.ac.in

subjects are shaped by a variety of disciplines, such as computer science, geography, Geographic Information Science (GIS), and urban studies. This chapter's goals are to outline important ideas related to the developing area of geo-intelligence, explain the distinctions between geo-intelligence and standard artificial intelligence, combining GIS with AI, introducing the key of GeoAI technologies and applications.

Furthermore, its goal is to outline the essential components of a successful geographic data analysis; the principal use cases and model domains; the primary benefits and drawbacks of implementing GeoAI techniques. Additionally, as a case study, we provide a successful real-world application of GeoAI in urban areas.

Idea behind GeoAI

Due to the nature of the topic and how artificial intelligence and geospatial technologies intersect with several other scientific fields and subjects, it is necessary to first define some basic terms related to these technologies (Hu et al. 2019, Döllner 2020).

1. The study and development of machines or computing systems capable of learning, reasoning, and foresight—tasks typically carried out by humans—is known as artificial intelligence. This allows items to function correctly in their surrounding environments.

2. *Machine Learning*: This area of artificial intelligence combines statistical and numerical optimisation approaches to create models from data without explicitly programming all model parameters or computational steps.

3. *Deep learning*: A sophisticated kind of machine learning in which artificial neural networks and algorithms inspired by the human brain analyse vast volumes of data to find patterns and forecast outcomes.

4. *Convolutional Neural Network*: CNNs are a subset of deep learning methods wherein convolution is used in place of standard matrix multiplication in at least one neural network layer.

5. *GeoAI*: This new field of study combines advances in high-performance computing, data mining, AI/ML techniques (deep learning), and spatial science to extract information from large amounts of spatially distributed data.

6. *Geospatial*: An all-encompassing word for information and related technologies with a spatial or locational component, frequently in relation to Earth.

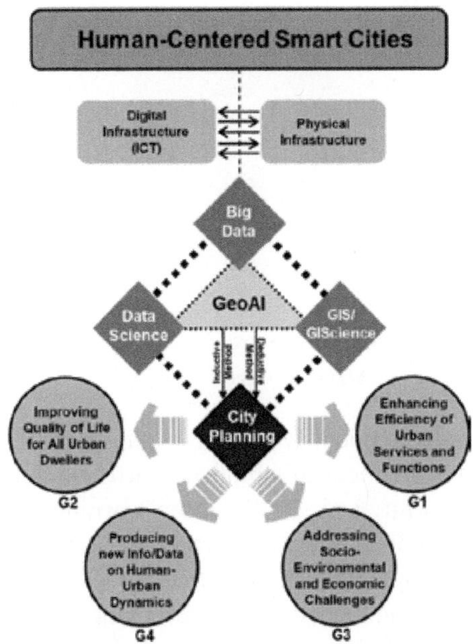

Figure 1. Visualization of Geo-AI infrastructure (Döllner 2020).

There are other meanings available for each of the concepts mentioned above, and each one was selected to be appropriate for the purposes of this investigation. These days, AI and other words from earlier are widely used. It is thus more important than ever to understand what it is and how it differs from others. Despite their possible close relationship, these words are not interchangeable. To visualise this, please refer to Figure 1.

GeoAI (Dong et al. 2023, VoPham et al. 2018) branch of study has been developed and uses high-performance computing, data mining, machine learning (including deep learning), and breakthroughs in spatial science to extract knowledge from spatial large data. GeoAI (Boulos et al. 2019), provides an example of how to merge elements of geographic information systems (GIS) and artificial intelligence (AI).

Thus, the intersection of artificial intelligence, photography, and geospatial technologies is known as GeoAI. Furthermore, another way to describe geospatial AI is as a new type of machine learning with a geographic foundation. Figure 2 presents the GeoAI idea.

The term artificial intelligence (AI) refers to devices that use task automation to generate scalable insights from massive data sets in order to understand the outside environment (Cresson 2019). A task is carried out by a computer that is relatively as intelligent as a person.

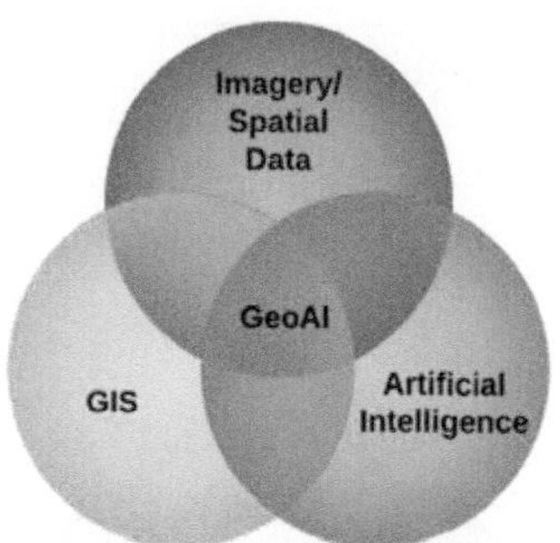

GeoAI = GIS + Artificial Intelligence

Figure 2. Concept of GeoAI.

Teaching computers to identify patterns in unprocessed data and repeatedly extract information is the focus of a branch of artificial intelligence known as "machine learning" (Zhu et al. 2017). Deep learning (DL) is considered a subgroup type of machine learning that draws inspiration from brain function. It is a powerful and adaptable technique that allows computers to learn from experience and understand the world as a nested hierarchy of concepts, where the computer is able to learn complicated concepts by building them from simpler concepts (Cresson 2019). Natural language processing, autonomous driving, and computer vision have all benefited from deep learning.

However, data mining refers to techniques for sorting through massive information in order to identify unique and fascinating patterns. One such technique is identifying recurring item groups in online transaction logs. A number of data mining techniques were developed as part of machine learning (Boulos et al. 2019). A new branch of research called GeoAI uses artificial intelligence (AI) techniques like deep mining, machine learning, and deep learning to extract useful information from massive volumes of geographical data. Comparable to this, GeoAI is a branch of spatial science that focuses on applying AI technologies to analyse vast amounts of geographic data so that decisions can be made and actions can be taken. Processing and analysing spatial data require the use of specialised spatial technologies, like GIS.

Distinctions between Standard AI and GeoAI

First, the logic needed to solve geospatial data and spatiotemporal data analysis differs greatly from other traditional AI tasks. Vector data, which shows how related objects are, must be correctly analysed using various types of algorithms, such as CNN.

Additionally, GeoAI is still a young field that requires creativity to solve a variety of issues, including hyperspectral imaging, real-time applications, and higher dimensional data. The most significant distinctions between GeoAI and

Table 1. Most significant distinction between GeoAI and traditional AI (Li et al. 2016).

Elements	Description
Data type	Geographical data is also dynamic since it pertains to the actual location and proximity of individuals and assets in both space and time.
Data volume	Because it is one of the data-intensive application areas where data are very large, complex, rapidly growing, and frequently include real-time elements, the volume of geospatial data is typically very high, which, frequently spanning decades, were recorded by satellites, devices, and sensors.
Higher dimensional data	Geospatial fields are using more high-dimensional data than traditional geometry-based analysis, which uses low-dimensional spaces. Both structured and unstructured geospatial data become more high-dimensional when combined with real-world data, increasing the complexity of AI techniques required to solve a particular problem.
Pre-processing for AI	More preprocessing is needed for geospatial data than for conventional AI models. The majority of conventional AI tools are not designed to comprehend ideas like terrain, spatial proximity, geospatial connectivity, etc.
Real-time applications	As opposed to text or images, which are analyzed by standard AI models, many GeoAI applications, such as autonomous vehicles, predictive routing, and asset tracking, require real-time processing and prompt results for decision making using complex and structurally different data.
Hyperspectral	Geospatial data, in contrast to simple images, can span several spectral frequencies that go beyond visible light frequencies. The term "multi-spectral" refers to data that includes both ultraviolet and infrared bands; the term "hyperspectral" refers to data that includes all available frequency bands. Different methods and domain knowledge are needed for hyperspectral data analysis because it is possible to identify the mineral components of objects that have been detected.

traditional AI are shown in Table 1. GeoAI is a vast multidisciplinary area that connects several fields including computer science, engineering, statistics, geography, and urban planning because of all these features of geospatial data. Some of the important applications employing Artificial Intelligence on spatial data can be accessed in literature (Chaurasia et al. 2021, Uniyal et al. 2021, Jena et al. 2021, Dixit et al. 2021, Anuranjan et al. 2022, Inampudi et al. 2020, Ayachit et al. 2020).

Combining AI and GIS

It is evident that fusing AI with GIS has proven to be extremely promising for a more productive and efficient future. AI took a significant stride that has been emerging on its own in GIS. The domains of GIS and AI have significantly converged through the past ten years. GIS is a highly regarded technology that offers enormous data sets and a broad range of AI applications (Kuldeep et al. 2017). Recent projects have also aimed to investigate several methods for distributional predictive modelling of geographic phenomena.

AI seeks to apply its methods in this way to advance intelligent information processing in GIS. Additionally, it provides great modelling solutions, including practical applications (Swati et al. 2017). The primary objectives of AI techniques in GIS applications are as follows:

- Enhance spatial pattern selection techniques
- Achieve prediction accuracy and assign spatial modelling techniques both separately and collectively for a variety of datasets

By factor sensitivity testing and rule extraction, artificial intelligence (AI) tools provide light on important spatial functions and processes.

Tools for GeoAI

The analysis of geodata has gained popularity due to the exponential development in the use of AI services. GeoAI systems offer tools that assist in all stages of the data science process, such as data preparation, exploratory data analysis, model training, geospatial analysis, and, at the end, findings dissemination through web layers and maps (Hu et al. 2018). A summary of ML's primary elements in GIS is shown in Figure 3.

With the use of distributed computing, AI has advanced to the point where it can now deploy trained models for feature extraction or classification, enabling the large-scale deployment of deep learning models. In addition, it offers resources to help with data preparation for DL procedures (Godinho et al. 2021).

GIS Software: In addition to examining geographical data, there are various types of GIS software packages that can be used to create, manage, analyse, and display data on a map.

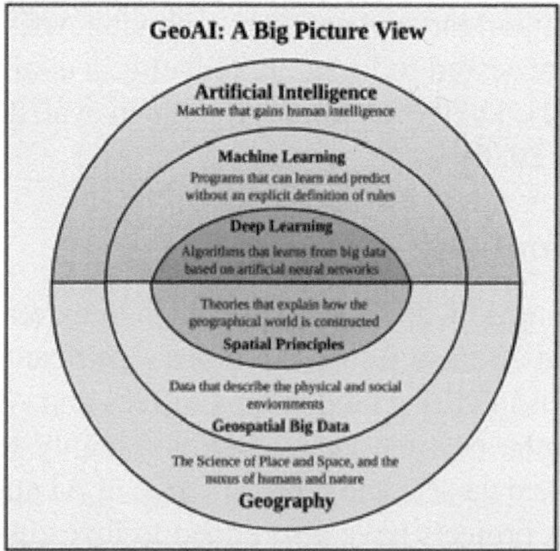

Figure 3. Concept of GeoAI (Hu et al. 2018).

Crucial Steps for Effective Analysis of Geospatial Data

First and foremost, it is critical to stress that spatial analysis requires a workflow and an ongoing approach to issue solving in the actual world. It cannot be accomplished simply by running software or a model. For geographic data analysis to be successful, the following stages are crucial (Katal et al. 2014).

- *Questions that define*: create suppositions and geographical inquiries
- *Examine the information*: assess the quality, completeness, and measurement constraints (resolution and scale) of the data to ascertain the supportable degree of analysis and interpretation
- *Examine and assess*: divide the issue into manageable, model-able components. Analyse and quantify the geographical questions
- *Analyse the outcomes*: assess and interpret the findings in the light of the issue raised; the constraints of the data; the correctness; and other ramifications
- *Repeat as necessary*: geospatial analysis is an ongoing, iterative process that frequently prompts new inquiries and advancements
- *Present the outcomes*: when the best data and analysis can be shared and presented to a wider audience in an efficient manner, their value increases
- *Make a decision*: GIS and geographical analysis are tools that assist in decision-making, and a good spatial analysis process frequently produces the knowledge required to motivate choices and actions

Important GeoAI Models and Application Domains

Advances in the use of AI to geospatial data are beginning to show encouraging signs of being able to unlock the real potential of this technology in today's competitive environment. Even though most large-scale projects are applied and explicitly geographic, there are currently not many AI techniques available. There is an increasing recognition of the importance of spatial diversity, encompassing the number and distribution of locations, along with the significance of temporal changes and their scales, in developing applications and models that facilitate research and decision-making (Lee 2017). In order to support innovation and uphold sustainable development, commercial and governmental entities can rely on the services provided by GeoAI techniques and systems.

Obstacles to the Adoption of GeoAI Techniques and Tools

It goes without saying that it is necessary to take into account the unique obstacles presented by working with geographical data for the correct deployment of GeoAI's methodology and procedures. Even though GeoAI holds promise for enhancing growth, efficiency, security, and other facets of life, there are some obstacles that can be summed up as follows:

- *Increasing hardware costs*: as geographical data continues to expand; more powerful computers will be needed to process it. As a result, the hardware needed to run such a system will become more expensive.

- *Violation of ethics*: there are ethical concerns around the collection of high-resolution geospatial data.

- *Lack of experts*: since GeoAI is a relatively new technology, the field lacks experts. Besides comprehending technology, one must also be able to examine and assess data (Boyd et al. 2011).

- *Quality of data*: the quality of the data is crucial to decision-making trust. When data comes from more sources and gets more unstructured, its quality usually declines. A data quality control process needs to be implemented in order to develop quality measures, analyse data quality, repair erroneous data, and assess the costs and benefits of quality assurance (Li et al. 2022).

- *Errors in the data*: knowing the source of the data is crucial to minimise errors caused by using several datasets. It is critical to avoid or lessen bias in data interpretation.

- *Data cleaning*: the process of eliminating redundant, inaccurate, and incomplete data from the source of input geographic data is known as data cleaning. It must be the first stage of any big data project that involves geography. Generally speaking, datasets contain a lot of redundant information that should be eliminated to reduce the overall cost of the project (Grace et al. 2020). The duration of geographical data for most application scenarios in GeoAI systems, is a must for real-time big geographic data analytics. The life cycle of the data should be described, with stated potential value, and the computing mechanism that makes the analytics process real-time tabled in order to add value to the analysis (Grace et al. 2020).

- *Security and privacy*: because data will be moving across several types of networks, GeoAI systems and applications need to have very high security and privacy. Databases also hold information and data that is confidential. Thus, in smart city applications, the greater concern is the high degree of security and privacy against unauthorised access (Li et al. 2022).

- *Geospatial data analytics system*: conventional RDBMS lack scalability and expandability and are only appropriate for structured data. Even when unstructured data is processed using geographic (non-relational) databases, there are issues with their functionality. To ensure flexibility, a system that combines the advantages of relational and non-relational database systems must be designed.

- *Geospatial data visualisation*: At every stage of the data analysis process, visualisation aids in decision making. Research on Online Analytics Processing (OLAP) and data warehousing still includes visualisation-related topics. Tools for high-dimensional data visualisation are available.

Benefits of GeoAI system

Vast geospatial datasets amassed from surveys, sensors, and social media offer wealthy information and geographic context for specific GeoAI blessings that may gift vast possibilities. These are the primary blessings of the GeoAI system. The maximum crucial ones are indexed below.

- The growing availability of geospatial information, the improvement of AI, and the provision of great computing electricity are accelerating the virtual use of geographic information.

- Geospatial information and AI era may be blended to supply extra powerful virtual transformation solutions.

- GeoAI virtual technology has the capability to boost up the virtual financial system and maximise its position in addressing local and international challenges.

- AI technology provides new possibilities to combine and leverage geospatial information and generate geospatial insights and predictions.

- GeoAI era may be used to enhance particular stages of the heterogeneous information lifecycle and maximise the blessings from those datasets with the aid of using assisted heterogeneous information at some point of the lifecycle of information structures. Additionally, fees can be reduced as performance and productivity grow.

- Significant problems arise in scaling GeoAI structures to address complicated information situations. In particular, it is possible to percentage the ones primarily based totally on deep learning that require huge quantities of labelled schooling information, version structure searches, etc. crushed with the aid of using unique possibilities in the geo domain.

- Combining human and device intelligence allows deeper expertise of spatial information. Common problems in deciphering spatial information are with robustness and reliability.

Case Study: By Automating Map Updates, GeoAI Improves Services for Residents in Urban Areas (Grace et al. 2020)

By using GIS to create a deep mastering version and schooling dataset, PACI is capable of automating a massive part of its mapping, which ends up in on the spot time and price financial savings in addition to development in information accuracy.

Principal advantages of Urban regions finder

- GeoAI Machine Learning Model permit it to routinely replace the base maps of urban regions with brand-new streets and structures.

- The time invested in developing and refining a deep mastering version produces ever-greater specific outcomes.

- Urban regions humans and citizens gain from time financial savings and accuracy improvements without delay, which allows them to attain higher information and steerage at a decreased price.

Major infrastructure expansion is anticipated under Urban areas Vision 2035, and with an 11% increase in infrastructure spending targeted under the Urban areas National Development Plan, transformation is already well under way. Building initiatives include the construction of the longest causeway in the world, a new passenger terminal for the airport and Silk City, a new residential neighbourhood with 500,000 residents. The nation must adapt to a landscape that is always changing. Urban areas Finder has to capture and portray this dynamic with an official, up-to-date geographical representation of the whole nation in order to maintain confidence.

Making the Transfer to Automation

When PACI released its GIS programme in 2011, it required a current, complete base map of Urban regions. The GIS crew first created Urban regions Finder in 2012 through compiling statistics from numerous ministries, inner paper maps from PACI, and AutoCAD documents to provide the prevailing base map. They used satellite TV for PC images to acquire statistics for the map with a purpose to replace the improvement of latest systems and adjustments to the city`s roadways. But the city infrastructure of urban regions had altered yet again by the time the team found the adjustments and re-entered them into the Urban regions Finder database.

Deep Learning to Know to Get the Greatest Answer

As visible in Figure 4 and Figure 5, PACI used system getting to know (ML) to extract constructing footprints and avenue facts from the satellite TV for PC image with the least quantity of human input. A state-of-the-art sort of system getting to know referred to as deep getting to know involves education, a laptop to locate styles in big volumes of facts, and the ability to discover and extract the information desired.

The primary duties of the GIS deep learning processing are displayed in Figure 4 and Figure 5.

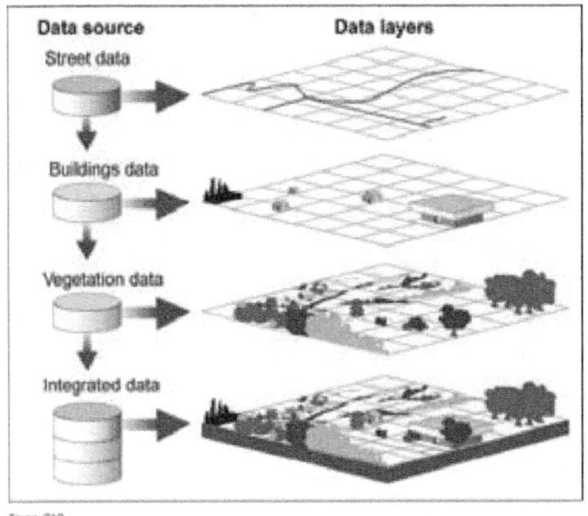

Figure 4. Visualization of buildings and highways (Grace et al. 2020).

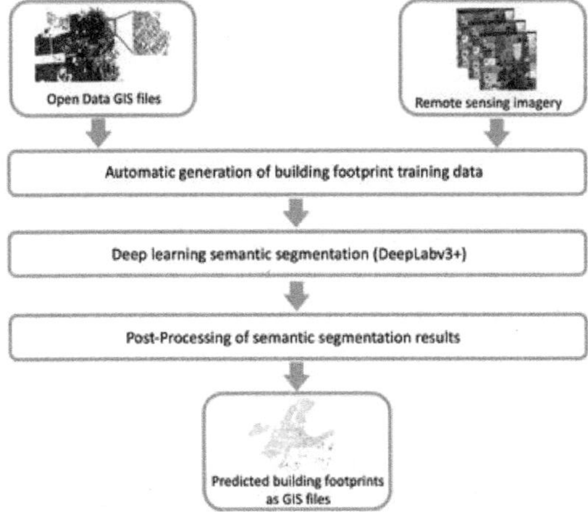

Figure 5. Deep learning system output (Grace et al. 2020).

When implemented correctly, the algorithm responds rapidly, comprehensively, and even detects changes that the human eye might overlook. The GIS team at PACI had to train the computer to identify building footprints from satellite data and identify which ones had changed since the last set of photos were taken.

Because PACI intended the computer to notify it of any new developments around the nation, its role was a little more complicated. The PACI team had to create a shared geographic framework that would cover the current database. In order to train the model to scan 3000 square kilometres of satellite footage, the crew was able to collect 75 square kilometres of data.

After the machine learning model was prepared, PACI spent time honing and perfecting it. As a result, the database can be updated in around three hours and even more accurate maps may be produced using the model and training data set. Furthermore, PACI may keep improving and changing this model in the future to account for newly developed building kinds, architectural styles, and other elements.

Achieving the Objective

Now, the model could do a work that used to require five individuals a year to finish. The computer was trained over time to successfully extract characteristics from satellite pictures and save them in the GIS repository.

The organisation is now confident that it can maintain Urban areas Finder and the Urban areas base map as the go-to sources of information in Urban areas by updating them on time. The tiny team at PACI will be able to continue developing to remain ahead of the competition because to the time and money savings from this automation. The knowledge gained by utilising deep learning and remotely sensed data will help inspire greater creativity and new ideas among PACI employees.

Suggestions

- It is necessary to concentrate on geospatial artificial intelligence (AI) rather than just hardware or the Internet of Things.
- Funding should be provided for basic geospatial artificial intelligence programmes.
- Improved long-term planning and strategy are needed for many applications of geospatial AI.
- It is imperative to strike a more equitable balance between immediate and long-term requirements for fundamental research, innovation, and development.

- Greater national financial assistance for academic geospatial artificial intelligence research.
- Improved technology transfer and company partnerships for the advancement of geographical artificial intelligence.
- A government push for artificial intelligence in business and other economic domains.
- Creation of a national agenda for geospatial AI research to support a long-term, well-coordinated effort. It is crucial for funding and building the geospatial data infrastructure that gives businesses and researchers access to high-quality data sets.
- Creation of systems to teach the general public how to use AI tools and techniques, both re- and upskilling them.
- An increase in national research funding to study the unique characteristics of both present and future GeoAI systems, as well as their wide-ranging effects on things like safety and health.
- Endorsement of research on the use of geospatial artificial intelligence (AI) in society that tackles and suggests new ways to maximise the benefits of technological advancements to society.
- Putting an emphasis on human skills to progress GeoAI research and development. More skilled individuals, such as students, from the data science field to focus on geospatial challenges may be enticed to join the program.

Conclusion

The following conclusions can be drawn from the above data mining, high performance computing, and intelligence (particularly machine learning and deep learning) to gain insights from large amounts of geographically relevant data. GeoAI systems have their origins in the field of geographic data and seek to expand and help organize the ways of thinking and methods for processing and evaluating vast amounts of spatial data. The purpose of this paper is to provide an overview of GeoAI technology, explain the concepts of GeoAI and how it differs from regular AI, and discuss the combination of AI with GIS and his GeoAI tools and software required for effective geographic information collection. It was about exploring possibilities. The most important GeoAI application areas and models as well as the advantages and disadvantages of using GeoAI techniques and technologies were also introduced. This chapter also provided a series of suggestions and case studies on the use of his GeoAI in urban areas. In summary, GeoAI is making remarkable and positive contributions to solving real-world problems across a variety of industries.

The aim of this study is to contribute to the current debate on the function of planning in smart cities the 21st century and to advocate a reinterpretation of the techno centric definition of smart cities. Achieve four key policy objectives: The goals of urban planning are:

1) Optimize the effectiveness of city services and city functions.

2) Improve the standard of living for all urban residents.

3) address pressing social, environmental and economic issues that can impact urban systems at different levels.

4) Supports the generation of spatial data, information, and understanding of human environment dynamics.

In this chapter, we demonstrate that our proposed human-centred GeoAI-powered urban planning framework can address both the issues of technical means and the socio-political, normative, and ethical issues of today's and future cities, we show that we can lay the foundations for an alternative vision of smart urbanism. This has been demonstrated through various empirical examples and research activities.

Furthermore, we believe that setting a clear vision and well-defined policy goals for planning will facilitate the convergence of different research initiatives and practices in this field. Indeed, the four policy objectives listed in this chapter support collaboration between planners from different disciplines.

Additionally, the research initiatives described in this chapter demonstrate how GeoAI has created rich opportunities for collaboration between scientists and practitioners in this field, and among planners and researchers in other fields. It shows. Future research may investigate the potential impact of his GeoAI on field integrity in planning.

References

Anuranjan, M. B., Divya Vani, C., Singh, C., Barman, S., Chaurasia, K. and Arun, P. V. 2022. Machine learning techniques for predicting dengue outbreak. pp. 45–56. *In*: Innovations in Information and Communication Technologies: Proceedings of ICIICT 2022. Singapore: Springer Nature Singapore.

Ayachit, S. S., Kumar, T., Deshpande, S., Sharma, N., Chaurasia, K. and Dixit, M. 2020, December. Predicting h1n1 and seasonal flu: Vaccine cases using ensemble learning approach. pp. 172–176. *In*: 2020 2nd International Conference on Advances in Computing, Communication Control and Networking (ICACCCN). IEEE.

Boulos, M. N. K., Peng, G. and Vopham, T. 2019. An overview of GeoAI applications in health and healthcare. International Journal of Health Geographics 18: 1–9. https://doi.org/10.1186/s12942-019-0171-2.

Boulos, M., Peng, G. and VoPham, T. 2019. An overview of GeoAI applications in health and healthcare. International Journal of Health Geographics 18(1). https://doi.org/10.1186/s12942-019-0171-2.

Boyd, D. and Crawford, K. 2011. Six provocations for Big Data. *In*: A Decade in Internet Time: Symposium on the Dynamics of the Internet and Society. https://papers.ssrn.com/sol3/papers.cfm?abstract_id=1926431.

Chaurasia, K., Dixit, M., Goyal, A., Uthej, K., Adhithyaram, S., Soni, A. et al. 2021, November. Deep learning based water feature mapping using Sentinel-2 satellite image. pp. 98–106. In SPIE Future Sensing Technologies 2021 (Vol. 11914).

Cresson, R. 2019. A framework for remote sensing images processing using deep learning techniques. IEEE Geoscience and Remote Sensing Letters 16(1): 25–29. https://doi.org/10.1109/lgrs.2018.2867949.

Dixit, M., Chaurasia, K. and Mishra, V. K. 2021. Automatic building extraction from high-resolution satellite images using deep learning techniques. pp. 773–783. In Proceedings of the International Conference on Paradigms of Computing, Communication and Data Sciences: PCCDS 2020. Springer Singapore.

Döllner, J. 2020. Geospatial artificial intelligence: potentials of machine learning for 3d point clouds and geospatial digital twins. PFG – Journal of Photogrammetry Remote Sensing and Geoinformation Science 88(1): 15–24. https://doi.org/10.1007/s41064-020-00102-3.

Dong, W., Motairek, I., Nasir, K., Chen, Z., Kim, U., Khalifa, Y. et al. 2023. Risk factors and geographic disparities in premature cardiovascular mortality in us counties: a machine learning approach. Scientific Reports 13(1). https://doi.org/10.1038/s41598-023-30188-9.

Ferneda, E., Cruz, F., Prado, H., Guadagnin, R., Santos, L., Santos, D. et al. 2016. Potential of ontology for interoperability in e-government. Brazilian Journal of Information Science 10(2). https://doi.org/10.36311/1981-1640.2016.v10n2.06.p47.

Godinho, I., Flores, C. and Marques, N. 2021. Consultation on the white paper on artificial intelligence—a European approach. Ulp Law Review 14(1): 157–167. https://doi.org/10.46294/ulplr-rdulp.v14i1.7475.

Grace, M., Scott, A., Sadler, J., Proverbs, D. and Grayson, N. 2020. Exploring the smart-natural city interface; re-imagining and re-integrating urban planning and governance. Emerald Open Research 2: 7. https://doi.org/10.35241/emeraldopenres.13226.1.

Hu, S., Karna, B. and Hildebrandt, K. 2018. Web-based multimedia mapping for spatial analysis and visualization in the digital humanities: a case study of language documentation in Nepal. Journal of Geovisualization and Spatial Analysis 2(1). https://doi.org/10.1007/s41651-017-0012-4.

Hu, Y., Li, W., Wright, D., Aydin, O., Wilson, D., Maher, O. et al. 2019. Artificial intelligence approaches. *In*: Wilson, J. P. (ed.). The Geographic Information Science & Technology Body of Knowledge, Association of American Geographers, Philadelphia. https://doi.org/10.22224/gistbok/2019.3.4.

Inampudi, S., Johnson, G., Jhaveri, J., Niranjan, S., Chaurasia, K. and Dixit, M. 2021. Machine learning based prediction of H1N1 and seasonal flu vaccination. pp. 139–150. *In*: Advanced Computing: 10th International Conference, IACC 2020, Panaji, Goa, India, December 5–6, 2020, Revised Selected Papers, Part I 10. Springer Singapore.

Jena, A. K., Potru, S. S., Balaji, D. R., Madu, A. and Chaurasia, K. 2021, April. Disaster risk mapping from aerial imagery using deep learning techniques. pp. 319–329. In International Conference on Unmanned Aerial System in Geomatics. Cham: Springer International Publishing.

Katal, A., Wazid, M. and Goudar, R. H. 2013. Big data: Issues, Challenges, Tools and Good Practices. 2013 Sixth International Conference on Contemporary Computing (IC3), Noida, 8–10 August 2013, 404–409. https://doi.org/10.1109/IC3.2013.6612229.

Kour, S., Dawn, A., Saha, S. and Biswas, B. 2023. Report on "the second united nations world geospatial information congress (un-wgic) pre-event". National Academy Science Letters 46(3): 263–270. https://doi.org/10.1007/s40009-023-01209-y.

Kuldeep, Banu, V., Uniyal, S. and Nagaraja, R. 2017. Space based inputs for health service development planning in rural areas using GIS. Geodesy and Cartography 43(1): 28–34.

Lee, I. 2017. Big Data: dimensions, evolution, impacts, and challenges. Business Horizons 60: 293–303. https://doi.org/10.1016/j.bushor.2017.01.004.

Li, W. and Hsu, C.-Y. 2022. GeoAI for large-scale image analysis and machine vision: recent progress of artificial intelligence in geography. ISPRS Int. J. Geo-Inf. 11: 385. https://doi.org/10.3390/ijgi11070385.

Swati, U., Sapanaa, C., Ravi, S., Babu, K. M., Vijaya, B., Satyanarayana, P. et al. 2017. Geospatial approach to identify gap areas for health facilities development in rural areas. ASCI Journal of Management 46.

Uniyal, S., Chaurasia, K., Purohit, S., Rao, S. S. and Mahammood, V. 2021, December. Geo-ML enabled above ground biomass and carbon estimation for urban forests. pp. 599–617. *In*: International Advanced Computing Conference. Cham: Springer International Publishing.

VoPham, T., Hart, J., Laden, F. and Chiang, Y. 2018. Emerging trends in geospatial artificial intelligence (GeoAI): potential applications for environmental epidemiology. Environmental Health 17(1). https://doi.org/10.1186/s12940-018-0386-x.

Zhu, X., Tuia, D., Mou, L., Xia, G., Zhang, L., Xu, F. et al. 2017. Deep learning in remote sensing: a comprehensive review and list of resources. IEEE geoscience and remote sensing magazine 5(4): 8–36.

10

Spectral Unmixing for Pollution Assessment in Water Bodies

Soorya Suresh,[1,]* *Arun P. V.*[1] and *Alok Porwal*[2]

Introduction

This study investigates advanced machine learning (ML) techniques to develop innovative approaches for processing and analyzing hyperspectral data obtained from satellite sensors observing Earth's water bodies. The complex pixel spectra in such data represent a mixture of constituent materials, including dissolved organic matter, algae, sediments, and pollutants. Several spectral unmixing approaches have been explored to identify these components and estimate their fractional abundance within each pixel. However, in the context of aquatic environments, there are unique challenges that must be addressed.

Firstly, Dynamic variability of aquatic constituents. Unlike the relatively stable mineral compositions on the water bodies, their constituents exhibit high temporal and spatial variability. Seasonal changes, weather events, and anthropogenic activities can rapidly alter the concentrations and distributions of various materials within a water body. Next, the presence of complex organic mixtures. In contrast, the well-defined spectral signatures of distinct minerals, pollutants, and other organic materials often display overlapping and convoluted spectral features, making their identification and quantification more challenging. The influence of atmospheric and water column effects is significant in aquatic hyperspectral data. These effects include atmospheric

[1] Indian Institute of Information Technology, Sri City, Tirupati, Andhra Pradesh.
[2] Indian Institute of Technology Bombay, Maharashtra.
Emails: arun.pv@iiits.in; alok.porwal@gmail.com
* Corresponding author: soorya.s@iiits.in

scattering, absorption by water molecules, and scattering within the water column.

They require robust correction techniques to accurately recover the inherent spectral signatures of waterborne materials. Traditional water quality monitoring often relies on ground truth data for training advanced analysis models. However, these methods can be limited by logistical challenges, cost constraints, and the lack of readily available ground truth data for diverse waterbodies. This chapter proposes a novel approach that overcomes these limitations by leveraging the power of hyperspectral imagery and unsupervised endmember extraction techniques. Instead of analyzing the entire image, we employ a masking technique to focus solely on the waterbody region. This eliminates the influence of land, clouds, and other irrelevant elements, improving the accuracy and specificity of our analysis. Imagine a satellite image capturing a coastline with a polluted river flowing into the ocean. Our masking technique would isolate the river portion, excluding the land areas and the vast expanse of the ocean, allowing us to concentrate on the pollutant signatures within the river itself. Due to the limited availability of ground truth data for specific pollutants in various waterbodies, we utilize unsupervised endmember extraction algorithms. These algorithms, such as Vertex Component Analysis (VCA) (Shimabukuro 1991), True Endmembers, and random pixel selection, can identify statistically significant spectral components within the masked waterbody image without prior knowledge of the specific pollutants present. This is akin to sifting through a collection of musical instruments and identifying distinct sound signatures, even if you don't know the names of the instruments themselves.

Once we have identified the endmembers using the aforementioned algorithms, we perform spectral unmixing using Fully Constrained Least Square Unmixing (FCLSU) (Shimabukuro 1991) and Convolutional Neural Network Autoencoder (CNNAEU) (Palsson 2020). This technique mathematically separates the mixed pixel spectra into the contributions of each identified endmember, revealing their fractional abundances within each pixel. By analogy, imagine each pixel as a mixed paint sample, and spectral unmixing as the process of separating the individual colors used to create that mixture. By analyzing the spectral characteristics and spatial distribution of the extracted endmembers, we can make inferences about the potential types and concentrations of pollutants present in the water body. This allows us to create pollution maps, highlighting areas with higher concentrations of specific pollutants and providing valuable insights for environmental monitoring and management efforts.

This approach represents a significant advancement in water quality assessment using hyperspectral imagery. By leveraging masking, unsupervised endmember extraction, and spectral unmixing, we can gain valuable insights into the presence and distribution of pollutants in waterbodies, even in the absence of extensive ground truth data. This paves the way for more efficient and effective water quality monitoring and pollution management strategies, contributing to a cleaner and healthier aquatic environment.

Study Area and Data Used

The main objective of this report is to devise novel techniques for spectral unmixing to characterize the compositional changes and geological context within water bodies. To achieve this objective, we have chosen Sentinel-2 (Figure 1), a part of the Copernicus Programme for Earth observation. This program systematically acquires optical images at high spatial resolution, ranging from 10 to 60 meters, covering both land and coastal waters. The Sentinel-2 (C. Sentinel-2 2020) mission is currently operated with two satellites, Sentinel-2A and Sentinel-2B, and a third satellite, Sentinel-2C, is in the testing phase, expected to launch in 2024. This mission supports a range of applications such as monitoring agriculture, managing emergencies (Dixit et al. 2021), classifying land cover (Chaurasia et al. 2020), and evaluating water quality. The European Space Agency oversees the development and operations of Sentinel-2, and a consortium led by Airbus Defence and Space in Friedrichshafen is responsible for manufacturing the satellites.

Figure 1. Image collected by Multispectral Instrument sensor onboard Sentinel-2A and Sentinel-2B platforms.

Overall Methodology

This study proposes a novel approach for spectral unmixing of hyperspectral imagery obtained from satellite sensors observing Earth's water bodies (Figure 2). It addresses the unique challenges presented by waterbody environments, with a focus on pollutant identification and abundance estimation in the absence of ground truth data.

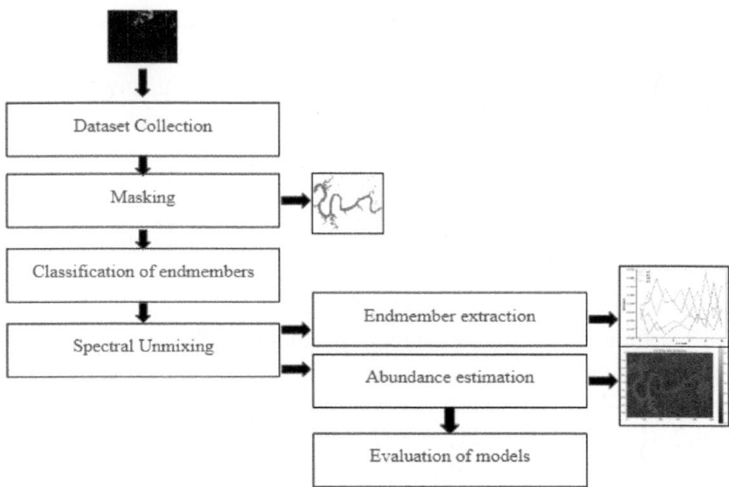

Figure 2. Proposed spectral unmixing technique to extract pollutants from an image collected by Multispectral Instrument sensor onboard Sentinel-2A and Sentinel-2B platforms.

To initiate the study, appropriate hyperspectral datasets relevant to the waterbody under consideration need to be selected. These datasets may include imagery from sensors such as Sentinel-2A, Sentinel-2B, Hyperion, or other sensors with suitable spectral resolution and coverage. Subsequently, preprocessing steps are applied to prepare the images for analysis. This may involve atmospheric correction, radiometric calibration, and geometric correction to ensure accurate spectral and spatial representation. A critical aspect of the proposed methodology involves implementing a masking technique to isolate the waterbody region from the surrounding land, clouds, and other irrelevant elements. This enhances the specificity of the analysis by focusing on the area of interest. Various masking techniques can be employed, including thresholding based on reflectance values, supervised or unsupervised image segmentation, or machine learning-based approaches.

Due to the lack of ground truth data for specific pollutants in various water bodies, the methodology advocates for the use of unsupervised endmember extraction algorithms. Techniques such as Virtual Component Analysis (VCA), True Endmembers, and random pixel selection can identify statistically significant spectral components within the masked waterbody image.

Following the extraction of endmembers, spectral unmixing is performed on the waterbody image. Popular choices include linear mixing models like Fully Constrained Least Squares (FCLS) and Convolutional Neural Network Autoencoder (CNNAEU). This step aims to estimate the fractional abundance of each identified endmember within each pixel, revealing their spatial distribution and relative concentrations within the waterbody.

The spectral characteristics and spatial distribution of the extracted endmembers are then analyzed to infer the potential types and concentrations of pollutants present in the water body. This interpretation involves relating endmember spectra to known spectral signatures of different pollutants and environmental parameters. The final step is to generate pollutant maps by visualizing the abundance and spatial distribution of identified pollutants based on the unmixing results. These maps provide valuable insights for environmental monitoring and pollution management efforts.

To assess the performance of the employed methods, researchers are encouraged to use quantitative metrics such as spectral angle mapper (SAM) and root mean squared error (RMSE). Additionally, there is an opportunity to explore the potential for classifying waterbody types or pollutant levels based on the extracted endmembers and unmixing results, potentially employing machine learning techniques like supervised classification algorithms.

By following this comprehensive methodology, researchers can leverage spectral unmixing techniques to gain valuable insights into the presence and distribution of pollutants in water bodies, even with limited ground truth data. This approach holds promise for advancing water quality monitoring and contributing to effective pollution management strategies for cleaner and healthier aquatic environments.

Data Preprocessing

In our waterbody analysis, the preprocessing stage tackles the inherent challenges of hyperspectral data volume and dimensionality. First, we combine the multiple band images (Figure 3) into a single data cube with dimensions (866, 1017, 11) representing height, width, and spectral bands Figure 5. However, instead of processing the entire image, we leverage a waterbody masking technique to extract only the relevant aquatic region Figure 4, significantly reducing computational complexity and focusing analysis on the area of interest. This masked data cube then undergoes principal component analysis as the dimensionality reduction technique, to retain key information while minimizing data size and simplifying subsequent processing steps. By efficiently pre-processing the data through masking and dimensionality reduction, we prepare a more manageable and informative representation for accurate spectral unmixing and pollutant analysis in the water bodies.

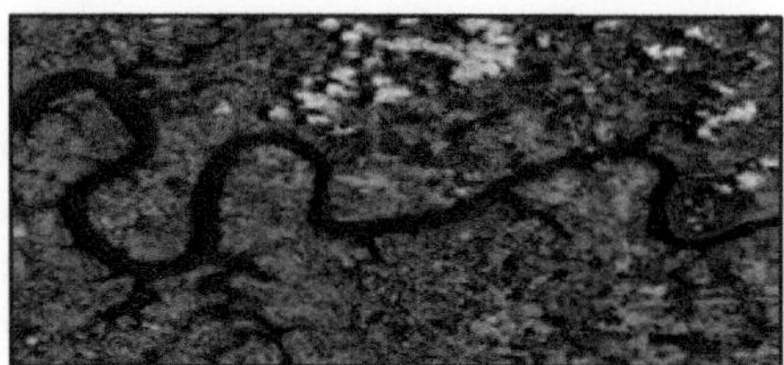

Figure 3. RGB composite of the image.

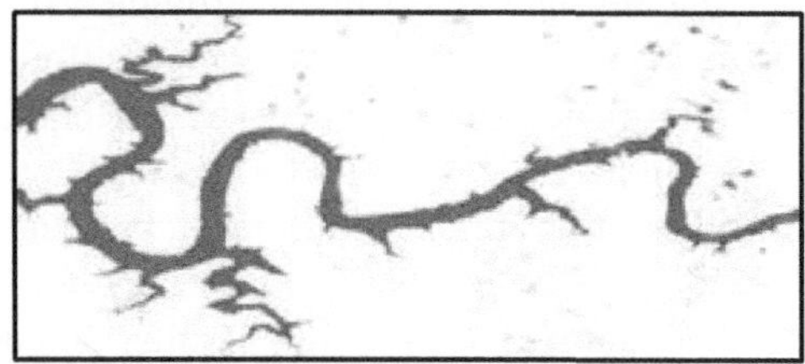

Figure 4. Masked image of water on image.

Spectral Unmixing

Spectral unmixing is a process used in hyperspectral image analysis to estimate the individual spectral signatures or endmembers present within a mixed pixel. It aims to determine the fractions of abundance of each end member in the pixel, providing valuable information about the composite and spatial distribution of different materials or substances within the image.

Endmember extraction methods

VCA (Vertex Component Analysis)

This method functions as an unsupervised technique to extract endmembers, operating based on the premise of linear spectral unmixing. It assumes that the signature of each pixel is an outcome of the linear combination of the spectra of endmembers existing in the scene. VCA operates based on two fundamental principles: initially, it associates endmembers with the vertices of simplexes, and secondly, it recognizes that the affine transformation of each simplex also results in a simplex. The process begins by assuming the presence of endmembers in the dataset and subsequently projecting data sample vectors iteratively onto directions perpendicular to the subspace covered by the previously identified hyperspectral endmembers. The new endmembers are related to extreme projections, both minimal and maximal. The iterative process continues until the predetermined range of endmembers is reached. The step-by-step procedure of the VCA algorithm is described as

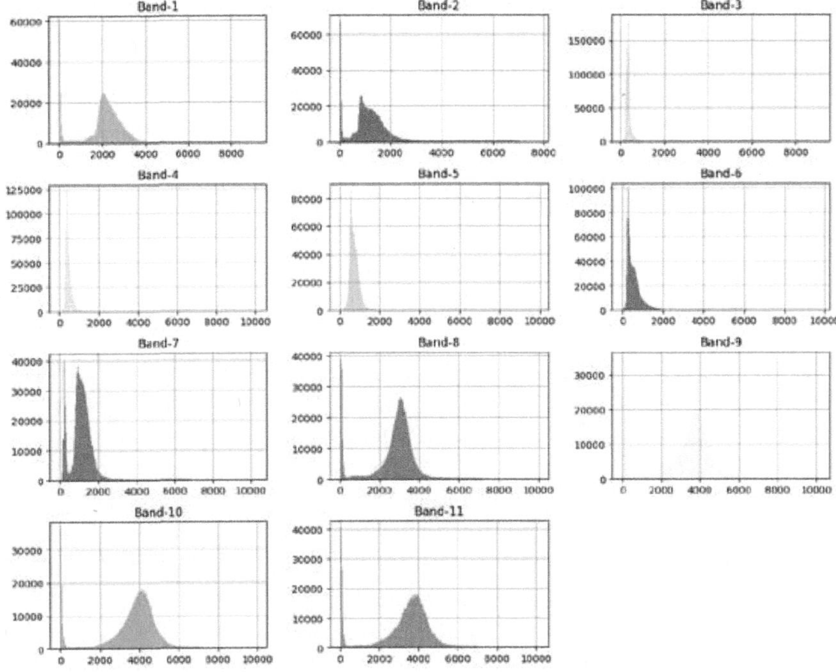

Figure 5. Plotting each band of the image.

follows: Let us define r as an L × 1 vector, where L denotes the total range of spectral channels.

$$r = e \cdot \alpha \tag{1}$$

In spectral unmixing, where e represents the abundance contribution of individual endmembers, α denotes the abundance fraction for each endmember, and p is the total count of endmembers within the image. Adhering to the non-negativity constraint, the abundance fractions of endmembers will meet the specified conditions outlined in equations (2) and (3):

$$0 <= \alpha_k <= 1 \tag{2}$$

$$\sum_{k=1}^{p} \alpha_k = 1 \tag{3}$$

In analyzing waterbody data, Vertex Component Analysis (VCA) shines as a powerful tool for unlocking the composition of each pixel. Imagine each pixel as a mixed paint sample, and VCA as the artistic detective separating the individual colors. It operates under the assumption that any pixel's spectrum is simply a blend of distinct spectral signatures called endmembers, like 'water blue,' 'algal green,' or 'pollutant red'. VCA starts by identifying some endmembers and then analyzes the remaining pixels for unique spectral components beyond those identified. This iterative process continues until all significant endmembers are revealed, painting a picture of the diverse substances contributing to each pixel's spectral fingerprint. By using VCA on masked waterbody data, we unmask the hidden composition of each pixel, providing valuable insights into the presence and distribution of pollutants within the waterbody, and paving the way for effective water quality monitoring and pollution management.

Abundance estimation

FCLSU (Fully Constrained Least Square Unmixing)

The primary goal of FCLSU is to find the proportions of individual endmembers within the spectrum of a mixed pixel, while adhering to physical constraints. FCLS, a linear unmixing technique, employs a simplex method to generate feasible solutions, ensuring the elimination of negative abundance values. Following the premise of a linear mixing model, FCLS interprets the spectrum of a mixed pixel by expressing it as a linear sum of endmember spectra. The abundance fractions, denoting the proportional contributions of individual endmembers, are constrained to be non-negative (according to Eq. 2) and collectively sum up to one (as per Eq. 3). FCLSU addresses a constrained optimization problem using the least squares method, aiming to minimize the disparity between the observed mixed pixel's spectrum and the calculated linear combination of endmember spectra. This optimization guarantees that the estimated abundances provide the best-fit solution, minimizing overall error. Notably, FCLS incorporates intrinsic constraints of non-negativity and summation to unity in the abundance estimation process, ensuring interpretability and physical relevance.

To extract endmembers from hyperspectral data, various methods such as Vertex VCA, True Endmembers, and Random Pixels are employed. The VCA algorithm treats endmembers as vertices of geometric shapes (simplexes) and projects the data iteratively onto directions that are orthogonal to the subspace spanned by the identified endmembers. This process isolates their contributions, and new endmembers are identified at the extremes of these projections, reflecting distinct spectral signatures. This iterative process continues until the desired number of endmembers is reached, revealing the underlying components of the data.

The True Endmembers method involves identifying endmember spectra in hyperspectral data using known or ground truth endmember spectra. This method leverages prior information about true endmembers, enhancing the extraction process. It entails comparing the spectral signatures of known endmembers with the spectral profiles of pixels in the hyperspectral data, using techniques like spectral angle mapping, spectral correlation, or least squares fitting to find the best matches. Abundance fractions of true endmembers in each pixel are estimated, aiding applications such as land cover classification. However, the effectiveness of this approach depends on the availability and reliability of true endmember data in real-world scenarios.

Another strategy for endmember extraction is Random Pixel Selection, which involves randomly selecting pixels from hyperspectral data and considering their spectra as potential endmembers. This approach aims to encompass data variability by including diverse pixel signatures. The process iteratively adds randomly selected pixels to the endmember set, attempting to capture a wide range of potential endmember spectra corresponding to different materials or classes within the scene. Despite its versatility, challenges related to noise and accuracy arise, as random selection may lead to less reliable endmembers. Therefore, careful consideration of the number of selected pixels, validation techniques, and inherent uncertainties is essential when employing this method for meaningful endmember extraction.

To estimate the proportion of each end member, the Fully Constrained least square Unmixing method (FCLSU) is used. It is a linear unmixing technique that employs a simplex method to generate feasible solutions, ensuring that negative abundance values are discarded. The equation for calculating abundance $\hat{A}$ is, Eq. 4.

$$(\hat{A}) = arg \ min_A \frac{1}{2} \|Y - EA\|_F^2 \tag{4}$$

$$(\hat{A}) = arg \ min_A \frac{1}{2} \|Y - EA\|_F^2 + \lambda R(A) \tag{5}$$

To assess pollutant abundances in water bodies, we present a model based on the principles of FCLSU. Initially, we preprocess the multi-band image data by masking out land areas, narrowing our focus to water regions, and reducing dimensionality for computational efficiency. Subsequently, we employ an innovative autoencoder approach to reconstruct the data, enhancing the accuracy of abundance estimation. The extraction of endmembers, which denote distinct spectral signatures, is performed using various methods such as Vertex Component Analysis (VCA) or true endmembers, if available. Finally, FCLSU computes the quantity of each endmember within each pixel, revealing the spatial distribution and relative concentrations of pollutants within the water body. This model adeptly unmixes mixed pixels, quantifies pollutant abundances, and offers valuable insights for water quality monitoring and pollutant management.

CNNAEU (Convolutional Neural Network Autoencoder)

An alternative model for assessing the fractional abundance of end members involves the utilization of CNNAEU (Convolutional Neural

Network-based Autoencoder). This innovative approach harnesses the capabilities of convolutional neural networks to delve into the intricate details embedded in both spectral and spatial data related to water bodies. The process begins by compressing the data through an autoencoder, extracting essential features that are pertinent to water characteristics. The resultant encoded representation then fuels CNNAEU, enabling it to uncover deeper spatial dependencies and perform unmixing of the data. This reveals the abundance of different water types within individual pixels. In contrast to conventional methods, CNNAEU stands out in capturing the nuanced interplay between spectral signatures and spatial patterns, providing a more precise depiction of the water-body landscape.

The method includes employing a CNN autoencoder trained on N patches, denoted as $Bi = \{x1, .., xn\}$, in which i ranges from 1 to N, and xj belongs to $RB \times 1$, originating from a Hyperspectral image (HSI), represented by $Y \in RD1 \times D2 \times B$. Notably, the method refrains from making use of biases, pooling layers, or upsampling. Rather, it exclusively utilizes non-strided convolution (CONV) layers that maintain the spatial resolution and input size throughout the procedure.

Figure 6 illustrates the schematic representation of the methodology. The initial layer serves as the input layer. Subsequently, the second layer, denoted as CONV 1, functions as a 2-D convolutional layer with 48 feature maps a 3×3 filter, and the activation function LeakyReLU. Batch normalization to enhance learning efficiency. Spatial dropout, a dropout variant tailored for CNNs aimed at curbing overfitting and enhancing generalization, is then applied with a dropout fee set at 0.2. This implies that entire feature maps undergo random zeroing with a 20% probability.

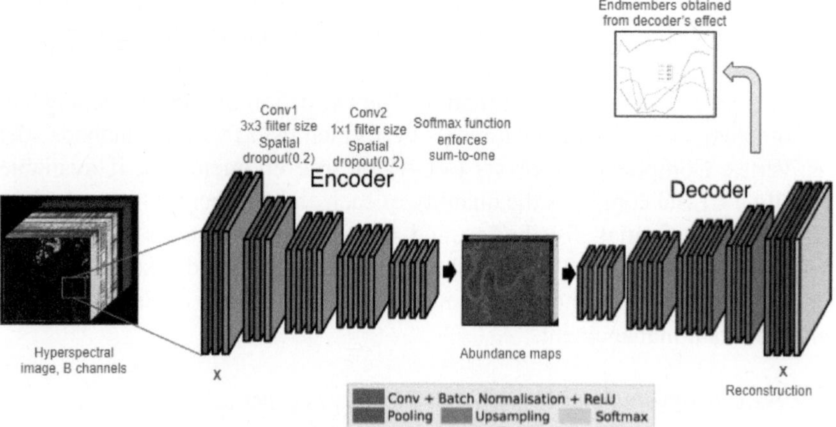

Figure 6. CNNAEU architecture.

Continuing the process, the subsequent convolutional layer, labeled as CONV 2, produces R feature maps using a 1×1 filter and employs LeakyReLU as the activation layer. Like the prior layer, the convolutional operation is succeeded by batch normalization and spatial dropout with a dropout rate of 0.2. This layer is responsible for generating the abundance maps. Moving forward, the final encoding phase of the autoencoder begins with a layer that enforces Additivity and Self-Coverage at every pixel. This involves applying the softmax function individually to the output feature maps array from the preceding layer, with the values scaled by a factor α (where $\alpha = 3.5$). The resulting feature maps portray abundances that collectively sum up to one at each pixel.

The final element in the system is the linear decoder layer. It operates as a convolutional layer with B feature maps, a filter size of $f \times f$ (where f is an odd number), and linear activation. Its primary function is to reconstruct the input patch using the abundance of patches acquired from the enforcing layer, which guarantees that the sum of abundances equals one. In the case of a filter size of 1×1, it can be illustrated that the filter's weights (W) generate a $B \times R$ matrix, where the columns signify the endmembers. For a filter size of $f \times f$, f^2 $B \times R$ matrices will exist. Furthermore, the formula for restoring pixel p in a patch can be expressed directly as follows:

$$\hat{x}P = \sum\nolimits_{m=1}^{f^2} W_m h_m \qquad (6)$$

The single index m is utilized to reference locations in the $f \times f$ patch. The autoencoder encodes the Hyperspectral Image (HSI), and the abundance maps are obtained from this encoding process. The endmembers are directly derived from the weights of the decoder layer. To assess the performance of the proposed CNNAEU method, we evaluate its effectiveness in the extraction of endmember and finding the quality of the abundance maps. The results are compared based on the abundance of the spectral Root Mean Squared Error, and the spectral angle distance.

It is important to highlight that while a lower Abundance Mean Absolute Error (MAE) indicates improved abundance estimation and a lower Spectral Root Mean Squared Error (RMSE) signifies enhanced reconstruction of the signal, a lower Relative Error (RE) doesn't necessarily equate to better abundance estimation. Spectral RMSE provides more informative insights into the performance evaluation. We will compare the RMSE and SAD values obtained from our model with those from different spectral unmixing models.

During the assessment of our approach, we give precedence to the excellence of the identified endmembers over the accuracy of the abundance maps. This preference is rooted in our method's tendency to generate abundance maps with a binary-like appearance, where the abundances predominantly lean towards being very low or very high, resembling maps used in classification. This behavior is shaped by factors such as batch normalization, the application of a ReLU-like activation, and the use of the softmax function to enforce the Abundance Sum-to-One Constraint (ASC). Another aspect to consider is the potential limitations tied to employing Spectral Angle Distance (SAD) loss for abundance estimation. While effective for endmember extraction, using SAD loss might not be the most suitable metric for data reconstruction due to its scale invariance. The adoption of SAD loss might introduce increased variability in abundance estimation, extending beyond what can be solely attributed to the variation in the quality of the extracted endmembers.

Results and Discussion

In the quest for optimal spectral unmixing of waterbodies, we explored a diverse set of techniques, including traditional methods like VCA, True-endmembers, and Random Pixel, alongside the deep learning approach of CNNAEU. Endmember extraction results (Figure 7) were evaluated using the SAD metric, revealing varying degrees of spectral accuracy. True-end

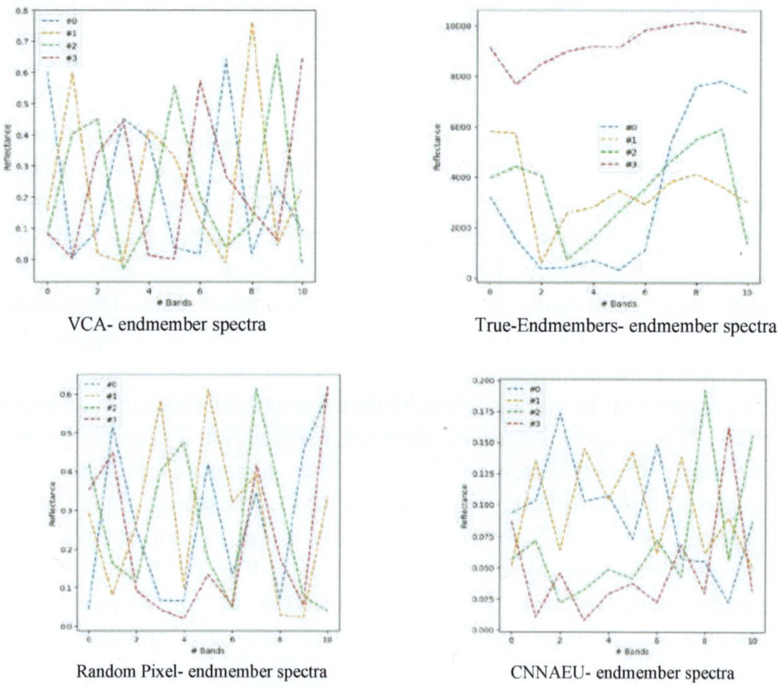

VCA- endmember spectra

True-Endmembers- endmember spectra

Random Pixel- endmember spectra

CNNAEU- endmember spectra

Figure 7. Endmember spectra.

Table 1. Evaluation of different models.

Dataset	Model	RMSE	SAD
Image collected by MSI (Multispectral Instrument) sensor onboard Sentinel-2A and Sentinel-2B platforms	VCA-FCLSU	18.10	48.33
	True endmembers- FCLSU	27.47	25.92
	Random Pixel- FCLSU	23.31	42.22
	CNNAEU	30.42	35.12

members demonstrated (Table 1) the lowest SAD value of 25.92, indicating the closest resemblance to reference members. However, regarding abundance estimation accuracy, measured using RMSE, VCA-FCLSU emerged as the frontrunner with an RMSE of 18.10, suggesting its superior ability to quantify endmember contributions within the water bodies dataset.

Notably, while CNNAEU exhibited a moderate SAD value of 35.12 and a higher RMSE of 30.42, its potential for refinement and adaptation to water bodies data remains promising. Further exploration of hyperparameter tuning, model architecture optimization, and integration of domain-specific knowledge could potentially enhance its performance in this context.

Conclusion

Our exploration of various spectral unmixing techniques for water bodies revealed valuable insights into their strengths and limitations. Traditional methods like VCA-FCLSU demonstrated superior abundance estimation accuracy, while True-end members excelled in capturing the pure spectral signatures of individual end members. The deep learning approach, CNNAEU, showed promise with its moderate performance in both aspects, suggesting potential for improvement with further optimization and adaptation to waterbody data. While VCA-FCLSU and True-endmembers currently take the lead in their respective areas, it is crucial to understand that the 'best' model ultimately depends on your research's specific goals and context. If precise endmember identification is paramount, True-endmembers might be the optimal choice. If accurate pixel-level abundance estimation is your priority, VCA-FCLSU could be the more suitable option. CNNAEU, with its flexibility and potential for refinement, presents a promising avenue for future exploration, particularly for researchers seeking a powerful and adaptable tool for waterbody spectral unmixing.

Future Scope

We will delve into datasets characterized by a higher number of spectral bands, enabling us to leverage more advanced spectral unmixing techniques.

By employing these deeper spectral unmixing methods, we aim to enhance our ability to extract pollutants from waterbodies with greater precision. This approach allows us to discern various types of pollutants more accurately, providing detailed insights into the composition of contaminants present in water environments.

Acknowledgments

This research was funded by the Indian Space Research Organization (ISRO), Dept. of Space, Govt. of India [Grant No. ISRO/SPO/GIA/2022]. The authors would also thank SSPO, ISRO, Prof. Alok Porwal, IIT Bombay, and Prof. Guneshwar T, NISER for the discussions and guidance.

References

Bioucas-Dias, J. M. and Nascimento, J. M. 2008. Hyperspectral subspace identification. IEEE Transactions on Geoscience and Remote Sensing 46(8): 2435–2445.

C. Sentinel-2. 2020. Sentinel online. Sensing 59(1): 535–549. Available: https://www.eoportal.org/satellite-missions/copernicus-sentinel-2.

Chaurasia, K., Neeraj, B., Burle, D. and Mishra, V. K. 2020, September. Topographical feature extraction using machine learning techniques from sentinel-2A imagery. pp. 1659–1662. In IGARSS 2020–2020 IEEE International Geoscience and Remote Sensing Symposium.

Chen, G. and Qian, S.-E. 2010. Denoising of hyperspectral imagery using principal component analysis and wavelet shrinkage. IEEE Transactions on Geoscience and Remote Sensing 49(3): 973–980.

Dixit, M., Chaurasia, K. and Mishra, V. K. 2021. Dilated-ResUnet: A novel deep learning architecture for building extraction from medium resolution multi-spectral satellite imagery. Expert Systems with Applications 184: 115530.

Palsson, B. M., Ulfarsson, O. and Sveinsson, J. R. 2020. Convolutional autoencoder for spectral–spatial hyperspectral unmixing. IEEE Transactions on Geoscience and Remote Sensing 59(1): 535–549.

Shimabukuro, Y. E. and Smith, J. A. 1991. The least-squares mixing models to generate fraction images derived from remote sensing multispectral data. IEEE Transactions on Geoscience and Remote Sensing 29(1): 16–20.

11

Assessing the Trend of Carbon Storage Changes from 1990 to 2020 Based on Land Use Changes

A Comparative Study between Hong Kong and Shenzhen

Jiongye Li[1] and *Yingwei Yan*[2,*]

Introduction

Research on carbon issues in urban contexts has been extensively explored using diverse methodologies. Many researchers focus on carbon emissions and sequestration related to energy consumption. For instance, Xia et al. (2023) categorized land use into several types, including industrial, transportation, and construction, estimating carbon emissions based on energy consumption specific to each land use type. Similarly, Chuai et al. (2015) calculated carbon emissions from four sources: energy consumption, industrial processes, waste, and natural vegetation, using data from the China Energy Statistical Yearbook. Sun and Xue (2020) expanded this approach by including variables like foreign trade, government intervention, and employment numbers in their carbon emission calculations. However, these methods assume the availability of detailed data, such as annual coal consumption or Gross Domestic Product (GDP), which might not be accessible in all regions, limiting their applicability.

[1] Department of Architecture, National University of Singapore, Singapore.

[2] Department of Geography, National University of Singapore, Singapore.

* Corresponding author: yingwei.yan@nus.edu.sg

Another approach to studying carbon emissions and sequestration relies on land use and types. This method estimates emissions based on carbon emission factors and the total area of each land use. Researchers can acquire land use data from agencies with well-documented records, such as the Data Center for Resources and Environmental Sciences, which provides Land Use and Land Cover (LULC) data for regions like Beijing, Tianjin, and Hebei (Wang et al. 2021). In cases where agencies do not provide LULC data, satellite imagery from sources like The National Aeronautics and Space Administration (NASA) can be a valuable tool, as demonstrated by Zhang et al. (2022).

Once LULC data is obtained, carbon emissions can be estimated. One method involves multiplying carbon emission factors by the area of each land use, as done by Zhang et al. (2022) and Rong et al. (2020). Another approach uses the Integrated Valuation of Ecosystem Services and Tradeoffs (InVEST) model from Stanford University's Natural Capital Project, which predicts carbon storage based on land use and carbon density indices. This model, applied by researchers like Zhu et al. (2019) and He et al. (2022), also requires carbon density indices for different land uses.

Economic development in China has led to significant land use changes in recent decades, driven by events like the 1978 Open Door Policy and China's 2001 World Trade Organization (WTO) accession. Specifically, Wu (2001) reports that foreign investment has flowed into mainland China, thereby accelerating the real estate market, and influencing urban development policies since the Open Door Policy. Furthermore, Wu (2001) highlights the significant role of investments from Hong Kong and Macau in boosting the urbanization rates of special economic zones, such as Shenzhen, Zhuhai, and Xiamen, following the policy. This pace of development was further accelerated when China joined the World Trade Organization (WTO) in 2001. According to Wu and Zhang (2022), the urban development of coastal regions, in particular, has been expedited by China's accession to the WTO Shenzhen, for example, transformed from a small village in 1979 into a mega-city by 2021 (Yu et al. 2019), experiencing a 3400% increase in developed land and a 44% loss in natural vegetation area from 1997 to 2017. In contrast, Hong Kong's urbanization rate has been less dramatic, with only a modest population increase from 1979 to 2020 and a slight rise in urbanization percentage (Wang et al. 2017). Given the differing developmental trajectories of Shenzhen and Hong Kong, this study aims to explore the spatiotemporal correlation between land use changes and carbon emissions at different time periods, a research field that demands stronger attention. This research addresses three key questions: (i) What are the changes in major land use types in Hong Kong and Shenzhen across decades? (ii) How have carbon emissions in these cities changed over the decades? (iii) What are the spatiotemporal trends in carbon emissions in both cities? The findings will offer insights for urban planners in balancing carbon emissions and urban development.

Data and Method

Data

This research primarily utilizes Landsat imagery, which is jointly managed by NASA and the U.S. Geological Survey (USGS). To acquire satellite images spanning from 1990 to the present, the study employs three types of Landsat imagery: Landsat 8, Landsat 7, and Landsat 5. Specifically, Landsat 8 imagery is employed for capturing data from the year 2020. Landsat 7 imagery is used to obtain satellite images for the years 2010 and 2000. Lastly, Landsat 5 imagery is utilized to gather data from the year 1990.

Method

This research methodology comprises three main components: land use classification for different time periods, calculation of carbon emissions and sequestration related to land use, and analysis of trends in carbon emissions and sequestration.

First, Google Earth Engine (GEE) is utilized to acquire Landsat imagery across different decades for the study area. During acquisition, GEE ensures that all satellite imagery has a cloud cover of less than 4% and employs a composite method to minimize cloud obstruction in the imagery. This process guarantees that the composited Landsat imagery is minimally affected by cloud cover. Previous research examining the correlation between land use and carbon emissions has typically classified land into five categories: built-up or constructed areas, natural vegetation (defined in some studies as forest or woodland), water bodies, cropland, and barren land (Yan et al. 2021, Yan et al. 2022, Liang et al. 2021). Subsequently, a training dataset, comprising sampled points across these land types (built-up area, crop, natural vegetation, water, and barren land), is established. A validation dataset, encompassing all land types within the entire research site, is also assembled. A machine learning model, specifically a random natural vegetation model with 300 trees, is trained using the training dataset. This model, which inputs spectral band information, aims to predict land types. After training, the model undergoes validation with the validation dataset, where true labels for each sample are provided. The validation utilizes 'Overall Accuracy' in GEE, calculating accuracy by dividing true positives by the total number of samples. Once the model achieves sufficient accuracy, it is applied to classify the entire research area.

For carbon emissions, this research calculates them based on carbon emission factors and the areas of each land use type. The indices for carbon emission factors related to cropland, natural vegetation, water, bare land, and urban areas are derived from prior studies (Fang et al. 2007, Rong et al. 2020, Zhang et al. 2022), using the following formula as shown in equation (1):

$$Total\ Carbon\ Emission = \sum(LU_i * CEF_i) \tag{1}$$

where LU_i represents land use types, and CEF_i represents carbon emission factors.

The trends in carbon emissions for Hong Kong and Shenzhen over the past three decades are analyzed using pixel-wise linear regression. This analysis examines the correlation of carbon emissions at each pixel across the raster datasets of each decade, spanning all three decades. The resulting raster illustrates the correlation between carbon emissions and time at each pixel. Positive correlations are indicated by positive numbers and vice versa. This analysis reveals the trends in carbon emissions for both Hong Kong and Shenzhen.

Result

Figure 1 presents the land use of Hong Kong in 1990, with classification accuracy in GEE reported at 98%. According to the classified land use, 81.1% of Hong Kong was covered by natural vegetation, 6.2% was urban land, 2.2% was cropland, 4.7% was barren land, and the remainder was water. It should be noted that the boundary of Hong Kong in this research is delineated by a red line; other boundary definitions may yield different percentages for each land use type. Figure 1 also shows Hong Kong's land use in 2000, with a GEE-calculated overall accuracy of 93%. In this year, 65.9% of the land was natural vegetation, 20.7% urban, 1% cropland, and 7.8% barren land. The land use map in Hong Kong for 2010 achieves an overall accuracy of 95% in GEE. The natural vegetation covered 68.4% of the area, urban land 24.7%, cropland 1.9%, and barren land 1.1%. The land use map in 2020 has an overall GEE accuracy of 87.6%, at this time period, the natural vegetation comprised 75.3% of the area, urban land 16.8%, cropland 0.9%, and barren land 1.5%. Notably, the natural vegetation area decreased by approximately 15% from 1990 to 2000, with a concurrent increase in urban land of 15%. However, between 2000 to 2010, the percentages of urban area and natural vegetation remained relatively stable. From 2010 to 2020, the natural vegetation area increased, and the urban area decreased by around 10%. For Shenzhen, as the classified images were trained concurrently with those of Hong Kong, their accuracy corresponds with that of Hong Kong for the respective time periods. In 1990, 79.5% of Shenzhen's area was natural vegetation, 2.4% built-up land, 0.8% cropland, and 10.6% barren land. By 2000, natural vegetation coverage had decreased to 46%, with urban areas increasing to 27.8%, cropland at 7.8%,

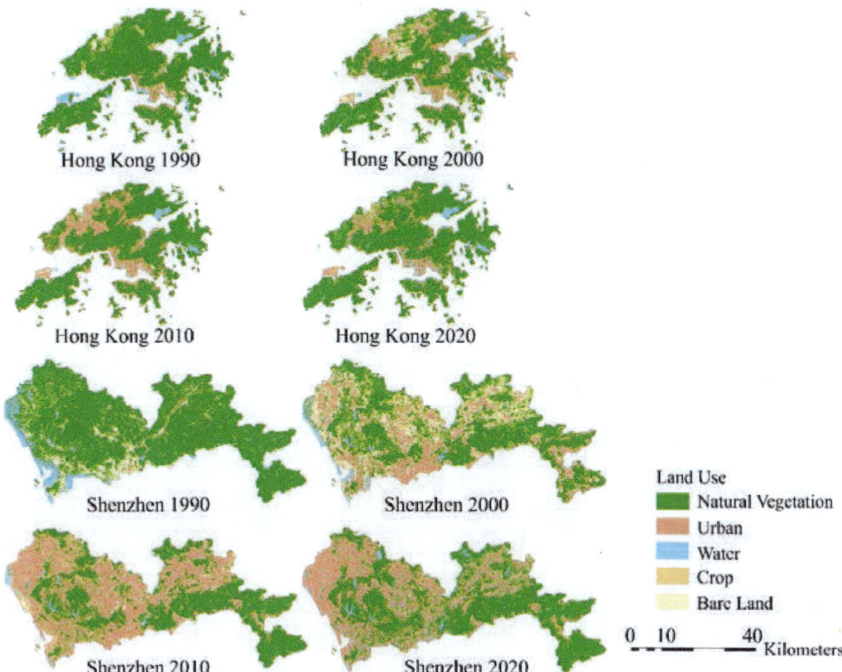

Figure 1. (Top) Land Use of Hong Kong from 1990 to 2020, and (Bottom) Land Use of Shenzhen from 1990 to 2020.

and barren land at 14.5%. In 2010, the natural vegetation constituted 43.4% of the area, urban land 45.6%, cropland 5.2%, and barren land 3.5%. In 2020, the natural vegetation area was 52.9%, urban land 40.4%, cropland 0.1%, and barren land 1.5%. From 1990 to 2000, there was a dramatic decrease in natural vegetation area and a significant increase in urban land.

Figure 2 displays the carbon emission and sequestration of Shenzhen across different time periods. The figure uses color coding, with redder areas indicating higher carbon emissions and greener areas signifying higher carbon sequestration. This comparison reveals a strong association between urban land expansion and increased carbon emissions from 1990 to 2010. In 1990, Shenzhen's carbon emission was 160,418,901.13 kg, which escalated to 2,998,023,241.90 kg in 2000 and 4,936,391,620.64 kg in 2010. By 2020, the emission was 4,363,483,298.05 kg.

This indicates an 18.7-fold increase from 1990 to 2000, a 65% increase from 2000 to 2010, and an 11.6% decrease from 2010 to 2020. In Hong Kong, carbon emissions were 332,590,908.83 kg in 1990, 1,256,622,727.86 kg in 2000, 1,506,664,930.10 kg in 2010, and 1,002,754,566.63 kg in 2020. The increase rate was 277.83% from 1990 to 2000, 19.90% from 2000 to 2010,

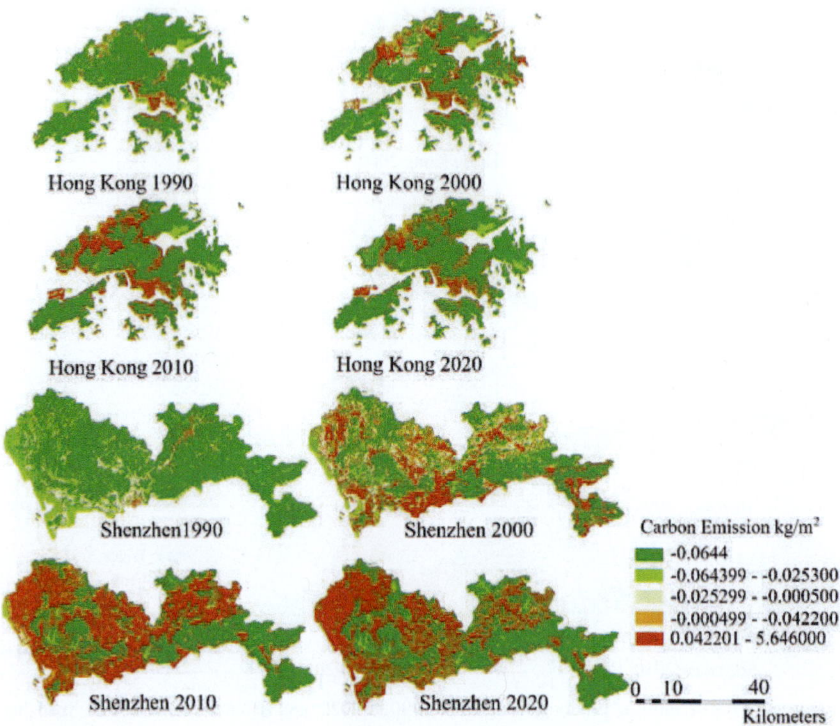

Figure 2. (Top) Carbon Emission of Hong Kong from 1990 to 2020, and (Bottom) Carbon Emission of Shenzhen from 1990 to 2020.

and a decrease of 33.45% from 2010 to 2020. Similar to Shenzhen, the most rapid increase occurred between 1990 and 2000, though the rate was not as high. From 2010 to 2020, both cities experienced a decrease in carbon emission rate.

The carbon emission trend diagram, Figure 3, with yellow indicating positive correlation and green indicating negative correlation between carbon emissions and years, shows that in Shenzhen, primarily the northwest and west regions saw an increase in carbon emissions from 1990 to 2020, whereas the southwest region exhibited a neutral correlation. In Hong Kong, mainly the northwest region is positively related to carbon emission.

Discussion

This research reveals that both Hong Kong and Shenzhen have undergone similar developmental processes. Between 1990 and 2000, both cities expanded their built-up land and significantly reduced natural vegetation land. However, Shenzhen experienced a more drastic urban growth during

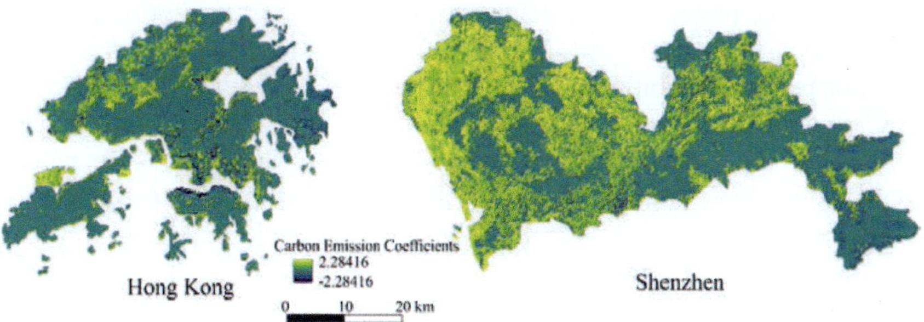

Figure 3. Correlation between carbon emission and Time of Hong Kong (Left) and Shenzhen (Right).

this period, with natural vegetation land decreasing from nearly 80% to 46%. In contrast, Hong Kong's natural vegetation land decreased by approximately 15% in the same timeframe. From 2000 to 2010, the natural vegetation coverage in both cities remained relatively unchanged from their 2000 levels. However, Shenzhen's urban land continued to expand significantly, increasing by about 20%, indicating a substantial conversion of other land types, such as cropland and barren land, into urban areas. Conversely, in Hong Kong, the percentage of bare land decreased by about 7%, with a modest urban land increase of around 4%. From 2010 to 2020, both Shenzhen and Hong Kong saw an increase in natural vegetation coverage and a decrease in urban land, suggesting a growing recognition of the importance of green spaces in urban environments.

In terms of carbon emissions over these three decades, the trends mirror the changes in land use. Both Shenzhen and Hong Kong experienced significant increases in carbon emissions from 1990 to 2000, with Shenzhen's emissions rising by approximately 18 times and Hong Kong's by almost 3 times. From 2000 to 2010, the rate of increase in carbon emissions slowed for both cities. Between 2010 and 2020, the carbon emissions in both cities decreased, correlating with the reduced urban land use and increased natural vegetation coverage.

Implications

The trend analysis of carbon emissions in Shenzhen and Hong Kong, conducted through linear regression mapping, reveals distinct patterns in different districts and areas. Shenzhen, Baoan, Guangming, and Longhua districts have consistently shown an increase in carbon emissions over the past few decades, indicating a positive correlation between carbon emissions and the passage of years. Conversely, in Dapeng and Yantian districts, the level of carbon emissions has remained constant throughout these years. In Hong Kong, areas displaying a positive correlation with carbon emissions over the same period include the northwest of the New Territories and parts of Kowloon.

For regions exhibiting a positive correlation between carbon emissions and time, it is recommended that mitigation measures be implemented. One effective strategy could be the reclamation of urban lands for natural vegetation, which would contribute to mitigating the upward trend in carbon emissions.

Limitations

Carbon emission factors, which quantify carbon emissions per unit area, are characterized by a range of units and values across various research papers, leading to inconsistencies and difficulties in comparing results across studies. For instance, Tian et al. (2022) defined carbon emission coefficients, also known as carbon density, in kilograms per square meter. In contrast, He et al. (2023) and He et al. (2016) expressed these coefficients in tons per hectare and megagrams per hectare, respectively. Furthermore, the values of carbon emission factors vary by location; for example, the aboveground carbon density in 'Grassland' is 0.7 Mg/hm² in Beijing, while it is 0.28 megagrams per hectare in Changchun, as reported by Li et al. (2022) and He et al. (2016). Despite these discrepancies, a consistent observation across multiple studies is the direct correlation between urban land and increased carbon emissions and an inverse correlation between natural vegetation land and carbon emissions. This observation is in accordance with the negative and positive signs of the carbon emission factors applied in our research. Thus, while the specific values of carbon emission factors in this study may slightly differ from those in other research, the directional trends and the relative magnitudes of carbon emission factors employed in our study are valid.

Conclusions

This study employs machine learning techniques to conduct a comparative analysis of land use and carbon emissions in Shenzhen and Hong Kong, focusing on emission trends over the past three decades. Specifically, the Random Forest algorithm, integrated with remote sensing data, has been applied on the GEE Platform. This approach enables the high-accuracy classification of land use across different time periods within the research area. Additionally, Linear Regression, in conjunction with remote sensing data, is utilized to analyze carbon emission trends at the pixel level in satellite imagery. This methodology offers a novel approach to understanding the dynamic relationship between land use and carbon emissions in these rapidly developing urban areas. Shenzhen, a city in mainland China, has transformed from a village to a mega city since the inception of the Open Door Policy in

1978. Hong Kong, by contrast, was already a modern, wealthy, and commercial metropolis by the 1970s. Through a comparative analysis of these two cities, it has been observed that Shenzhen underwent rapid land use changes from 1990 to 2000, with its natural vegetation area decreasing from nearly 80% to 46%, and urban land expanding by almost 20% from 2000 to 2010. From 2010 to 2020, Shenzhen saw an increase in natural vegetation cover and a decrease in urban land. Conversely, in Hong Kong, natural vegetation cover decreased by about 15%, and urban land increased by a similar percentage from 1990 to 2000. From 2000 to 2010, urban land in Hong Kong only increased by 4%. Similar to Shenzhen, from 2010 to 2020, Hong Kong experienced an increase in natural vegetation cover and a decrease in urban land. Regarding carbon emissions, Shenzhen's emissions increased approximately 18-fold, and Hong Kong's increased almost threefold from 1990 to 2020. From 2000 to 2010, both cities continued to see an increase in carbon emissions, albeit at a slower rate, and a decrease from 2010 to 2020. The research also identifies areas in Shenzhen and Hong Kong that have shown a positive correlation with carbon emissions over these three decades, highlighting the need for mitigation measures to curb emissions in these regions.

Acknowledgement

This research/project is supported by the Ministry of Education, Singapore, under the Academic Research Fund Tier 1 [FY2022-FRC2-009].

References

Chuai, X., Huang, X., Wang, W., Zhao, R., Zhang, M. and Wu, C. 2015. Land use, total carbon emissions change and low carbon land management in Coastal Jiangsu, China. Journal of Cleaner Production 103: 77–86.

Fang, J., Guo, Z., Piao, S. and Chen, A. 2007. Terrestrial vegetation carbon sinks in China, 1981–2000. Science in China Series D: Earth Sciences 50(10): 1341–1350.

He, C., Zhang, D., Huang, Q. and Zhao, Y. 2016. Assessing the potential impacts of urban expansion on regional carbon storage by linking the LUSD-urban and InVEST models. Environmental Modelling & Software 75: 44–58.

He, Y., Ma, J., Zhang, C. and Yang, H. 2023. Spatio-temporal evolution and prediction of carbon storage in Guilin based on FLUS and InVEST models. Remote Sensing 15: 1445.

He, Y., Xia, C., Shao, Z. and Zhao, J. 2022. The spatiotemporal evolution and prediction of carbon storage: a case study of urban agglomeration in China's Beijing-Tianjin-Hebei Region. Land, 11: 858.

Li, Y., Liu, Z., Li, S. and Li, X. 2022. Multi-scenario simulation analysis of land use and carbon storage changes in changchun city based on FLUS and InVEST model. Land 11: 647.

Liang, Y., Hashimoto, S. and Liu, L. 2021. Integrated assessment of land-use/land-cover dynamics on carbon storage services in the Loess Plateau of China from 1995 to 2050. Ecological Indicators 120: 106939. Elsevier BV.

Rong, T., Zhang, P., Jing, W., Zhang, Y., Li, Y., Yang, D. et al. 2020. Carbon dioxide emissions and their driving forces of land use change based on economic contributive coefficient (ECC) and ecological support coefficient (ESC) in the lower yellow river region (1995–2018). Energies 13(10): 2600.

Sun, C. and Xue, C. Q. 2020. Shennan Road and the modernization of Shenzhen architecture. Frontiers of Architectural Research 9(3): 437–449.

Tian, L., Tao, Y., Fu, W., Li, T., Ren, F. and Li, M. 2022. Dynamic simulation of land use/cover change and assessment of forest ecosystem carbon storage under climate change scenarios in Guangdong Province, China. Remote Sensing 14: 2330.

Wang, A., Chan, E. H., Yeung, S. C. and Han, J. 2017. Urban fringe land use transitions in Hong Kong: From new towns to new development areas. Procedia Engineering 198: 707–719.

Wang, C., Zhan, J., Zhang, F., Liu, W. and Twumasi-Ankrah, M. J. 2021. Analysis of urban carbon balance based on land use dynamics in the Beijing-Tianjin-Hebei region, China. Journal of Cleaner Production 281: 125138.

Wu, F. 2001. China's recent urban development in the process of land and housing marketisation and economic globalisation. Habitat International 25(3): 273–289.

Wu, F. and Zhang, F. 2022. Rethinking China's urban governance: The role of the state in neighbourhoods, cities and regions. Progress in Human Geography 46(3): 775–797.

Xia, C., Zhang, J., Zhao, J., Xue, F., Li, Q., Fang, K. et al. 2023. Exploring potential of urban land-use management on carbon emissions—A case of Hangzhou, China. Ecological Indicators 146: 109902.

Yan, H., Guo, X., Zhao, S. and Yang, H. 2022. Variation of net carbon emissions from land use change in the Beijing-Tianjin-Hebei region during 1990–2020. Land 11(7): 997.

Yan, H., Li, W., Yang, H., Guo, X., Liu, X. and Jia, W. 2021. Estimation of the rational range of ecological compensation to address land degradation in the Poverty Belt around Beijing and Tianjin, China. Land 10(12): 1383.

Yu, W., Zhang, Y., Zhou, W., Wang, W. and Tang, R. 2019. Urban expansion in Shenzhen since 1970s: A retrospect of change from a village to a megacity from the space. Physics and Chemistry of the Earth, Parts A/B/C 110: 21–30.

Zhang, C.-y., Zhao, L., Zhang, H., Chen, M.-n., Fang, R.-y., Yao, Y. et al. 2022. Spatial-temporal characteristics of carbon emissions from land use change in Yellow River Delta region, China. Ecological Indicators 136: 108623.

Zhang, S., Yang, P., Xia, J., Wang, W., Cai, W., Chen, N. et al. 2022. Land use/land cover prediction and analysis of the middle reaches of the Yangtze River under different scenarios. Science of The Total Environment 833: 155238.

Zhu, E., Deng, J., Zhou, M., Gan, M., Jiang, R., Wang, K. and Shahtahmassebi, A. 2019. Carbon emissions induced by land-use and land-cover change from 1970 to 2010 in Zhejiang, China. Science of The Total Environment 646: 930–939.

12

Harnessing AI in Decision Support Systems for Comprehensive Pollution Management

Gaganpreet Singh

Introduction

The growing environmental challenge

In the modern world, an unparalleled environmental disaster is occurring, characterized mostly by rising levels of pollution. This dilemma manifests itself in a variety of ways, from smog-filled skies over bustling urban areas to plastic-infested seas in our oceans. The evidence of this problem is both pervasive and worrisome, highlighting the critical need for comprehensive environmental management techniques (Arashpour 2023).

Environmental management has always depended on fundamental but often reactive strategies. While these traditional tactics served their intended purpose in the past, they are rapidly proving insufficient in the face of multifaceted and intricate environmental concerns. One of the key weaknesses of these techniques is their responsive nature. Instead of predicting and preventing environmental hazards (Smith 2013, Kuldeep et al. 2021), they frequently focus on post-mortem mitigation. Furthermore, these systems

CSRE, IIT Bombay, Powai 400076.
Email: gpsingh@iitb.ac.in

usually suffer from a shortage of real-time data integration, which makes it difficult to respond to new environmental concerns in a fast and effective manner. Furthermore, traditional environmental management approaches continue to face considerable challenges in processing and analyzing enormous volumes of heterogeneous environmental data (Huang et al. 2011).

The need for advanced solutions

Given these issues, enhanced, proactive, and data-driven strategies are urgently needed. This is where Artificial Intelligence (AI)'s transformational potential comes into play (Silvestro et al. 2022). AI, with its amazing ability to handle massive amounts of data, learn from patterns, and do predictive analysis, provides a ray of hope to combat these environmental concerns.

AI's data handling prowess is not limited to managing big amounts, but also to dealing with data diversity and complexity. Environmental data includes a wide range of factors, including atmospheric conditions, water quality indices, geographical data, and biological markers. Traditional approaches cannot compete with AI's capacity to filter through this enormous and varied data, uncover patterns and connections, and extract meaningful insights. Furthermore, based on past and present data, AI systems can be developed to forecast future trends and events (Chaurasia et al. 2020a, Jena et al. 2021, Chaurasia et al. 2019), allowing for a change from reactive to proactive management of the environment (Rayhan 2023).

In response to these serious environmental concerns, AI emerges not simply as a technological solution, but as an essential evolution in our methodical approach to environmental management. Its application could represent a paradigm change from conventional techniques, paving the way for improved efficiency, effectiveness, and environmentally conscious methods.

Introduction to AI-based Decision Support Systems

This chapter introduces AI-based Decision Support Systems (DSS) as groundbreaking solutions to the current environmental predicament. DSS are sophisticated, interactive software-based systems that assist decision-makers in compiling and utilizing information from a myriad of sources—raw data, documents, personal knowledge, business models—to identify and solve problems, as well as make informed decisions.

AI integration in DSS

When integrated with AI, DSS transcends the capabilities of traditional systems, offering a level of analysis, prediction, and strategic planning that is essential for comprehensive pollution management. The core strength of

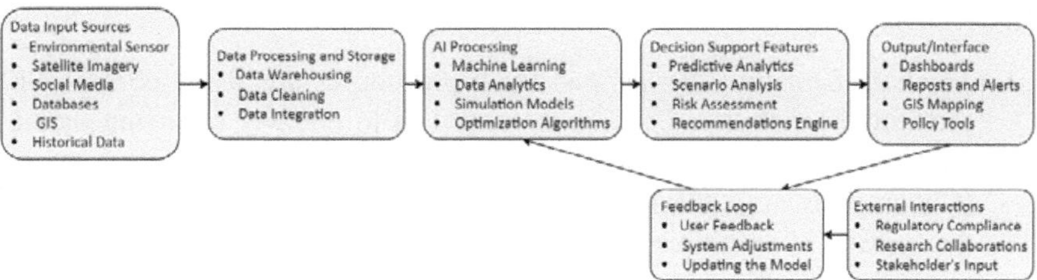

Figure 1. AI in DSS.

AI-based DSS lies in their ability to process and analyze large sets of data efficiently and effectively. This data can range from environmental readings, such as air and water quality indices, to socio-economic data, which are crucial for understanding the broader impacts of environmental policies (Wang et al. 2022).

Figure 1 illustrates the integration of AI algorithms with DSS, highlighting components such as data input, AI processing, and decision outputs.

AI integration allows DSS to not only handle the sheer volume of data but also to derive meaningful insights from it. Through advanced algorithms and learning models, AI systems within DSS can identify patterns, trends, and anomalies that might go unnoticed by human analysts. This capability is particularly crucial in environmental management, where early detection of potential issues can lead to timely and effective interventions.

Furthermore, AI-based DSS are provided with predictive analytics, which distinguishes them from traditional systems. Based on historical and present data trends, these systems can estimate future environmental conditions. This predictive power is critical for proactive environmental management because it enables stakeholders as well as policymakers to anticipate possible issues and undertake preventive measures.

The incorporation of AI into DSS also improves decision-making support. These systems may analyze complex events and make evidence-based suggestions, guiding users to the most effective and long-term solutions. This feature of AI-based DSS is especially useful in the realm of environmental management, where decisions frequently include complicated trade-offs and long-term consequences.

Leveraging AI in DSS

AI-based Decision Support Systems (DSS) leverage the predictive prowess of AI algorithms within a strategic framework, creating a synergetic solution that transcends traditional methods. This integration facilitates a nuanced understanding of environmental data, predictive modelling of pollution trends, and the development of effective mitigation strategies (Vinuesa et al. 2022).

AI Techniques in DSS

1. *Machine Learning Algorithms*: Such procedures are at the core of AI's predictive capability. They study past data to uncover recurring themes and patterns, which are then utilized to forecast possible future situations (Chaurasia et al. 2020a). In pollution control, this can involve projecting the growth in pollution levels depending on a variety of factors such as industrial operations, traffic flow, meteorological conditions, and more.

2. *Deep Learning Neural Networks*: They are a class of machine learning algorithms. These sophisticated networks imitate the structure and function of the human brain, allowing the processing of vast and complicated datasets. Deep learning can be used in environmental management to do tasks such as analyzing satellite photos to detect variations in land use or vegetation cover (Chaurasia et al. 2020b) that may indicate environmental degradation.

3. *Natural Language Processing (NLP)*: It is an artificial intelligence-based method that deals with computer-human interaction. These models are especially advantageous for processing and analyzing unstructured data like public opinions, news stories, and scientific literature. NLP has applications in environmental DSS to assess public attitude on pollution issues, analyze policy papers, and synthesize study findings.

Integrating AI into DSS

The integration of these AI techniques into DSS enables a multi-faceted approach to environmental management. By combining different AI modalities, these systems can provide a comprehensive analysis that includes not only hard data like pollution levels and environmental indicators but also soft data like public opinions and policy impacts. This holistic approach is crucial for developing effective and sustainable environmental strategies.

Objective of the Chapter

The primary objective of this chapter is to explore the role and potential of AI-based Decision Support Systems (DSS) in revolutionising the field of environmental management, with a particular focus on pollution control.

Role and potential of AI-based DSS

Data processing and analysis

One of the key roles of AI in DSS is the ability to process and analyze data from diverse sources. This includes satellite imagery for geospatial analysis, sensor data for real-time environmental monitoring, and social media feeds for public sentiment analysis.

Actionable insights and recommendations

The chapter will delve into how AI-based DSS can transform raw data into actionable insights. For instance, by analysing trends in air quality data, a DSS can recommend the best course of action to mitigate pollution levels.

Integration of GIS for enhanced decision-making

The importance of integrating Geographic Information Systems (GIS) with DSS will be highlighted. GIS provides spatial analysis capabilities that, when combined with AI-based DSS, can lead to more effective decision-making in pollution management.

Scope of the Chapter

The chapter will cover various aspects of AI-based DSS, including their architecture, functionalities, and real-world applications. It will discuss how AI techniques like machine learning and NLP can be utilized within DSS to address complex environmental challenges. The chapter will also explore the ethical considerations.

Structure of the Chapter

The chapter is structured to provide a comprehensive understanding of AI-based Decision Support Systems (DSS) in environmental management, with a focus on pollution control. The detailed structure is as follows:

AI applications in pollution mapping

This section explores how AI, particularly deep learning, is used in pollution mapping through the analysis of satellite imagery. The intricacies of different AI techniques like Convolutional Neural Networks (CNNs) in detecting and analysing pollution patterns from space are discussed.

Natural language processing for public sentiment analysis

The role of Natural Language Processing (NLP) in environmental contexts is examined, particularly in analysing public sentiment on environmental issues. The methodology and applications of NLP in processing discourse from various platforms like social media, forums, and news articles are explored.

Architecture of DSS and integration with GIS

The architecture of AI-based DSS is outlined, focusing on their core components and functionalities. The significance of integrating Geographic Information Systems (GIS) with DSS for enhanced spatial analysis and decision-making is highlighted.

Interdisciplinary collaboration and real-world applications

This section emphasizes the importance of interdisciplinary collaboration in the development and implementation of AI-powered DSS. It also showcases real-world applications of these systems in various environmental scenarios.

Ethical considerations and scalability of AI solutions

The chapter concludes with a discussion on the ethical considerations in the application of AI for environmental management and the scalability of AI solutions across different geographic and socio-economic contexts.

AI and Geospatial Technology in Pollution Mapping

Deep learning and satellite imagery

Understanding CNNs in the context of satellite imagery

Convolutional Neural Networks (CNNs) are a class of deep neural networks, primarily applied to analysing visual imagery. These networks are especially suited for data with a grid-like topology, such as images. CNNs consist of multiple layers of convolutions with learnable filters and pooling layers, which automatically and adaptively learn spatial hierarchies of features from input images.

CNNs are trained to identify several environmental signs that indicate pollution in satellite imagery analysis. This includes land use changes, vegetation health, the existence of toxic algal blooms in bodies of water, and anomalous discharges in thermal or infrared photography.

Processing and interpreting satellite data

The process begins with the acquisition of high-resolution satellite images, providing a comprehensive view of large geographical areas. The raw satellite data is then pre-processed to enhance quality, involving noise reduction, normalisation, and correction for atmospheric interference. The aim is to ensure the input data fed into the CNN is of the highest quality for accurate analysis.

CNNs automatically detect and extract relevant features from the pre-processed images, such as specific colour patterns, textures, or shapes indicative of pollution. These extracted features are then analyzed to identify pollution patterns. For instance, gradual changes in water body coloration could indicate chemical contamination or algal growth.

The final step involves pinpointing exact locations of hotspots where pollution levels are critically high and providing crucial information for environmental agencies and policymakers to take targeted actions.

Advantages and challenges

The use of CNNs for pollution mapping via satellite imagery offers several advantages, including scalability for large-scale monitoring; high accuracy in identifying pollution indicators; and cost-effectiveness compared to traditional monitoring methods. However, challenges such as the need for large datasets to train models and potential biases in data interpretation remain areas for future enhancement.

Natural Language Processing for Public Sentiment Analysis

The role of NLP in environmental context

Natural Language Processing (NLP) is critical for determining public sentiment on environmental concerns. Public opinion has become an extensive repository of data with the proliferation of social networks, discussion forums, and digital news platforms. NLP gives the capabilities required to evaluate and interpret this massive amount of unstructured data.

Analysing discourse on various platforms

Social media analysis

NLP algorithms sift through posts, tweets, and comments to extract data about environmental concerns. Sentiment analysis categorizes the emotional tone behind these posts as positive, negative, or neutral.

Forum discussions

Online forums feature detailed conversations about environmental issues. NLP techniques analyze these discussions to identify common themes, concerns, and the level of awareness among the public.

News article processing

NLP analyzes the frequency and context of environmental topics in the news, providing insights into media portrayal and its influence on public perception.

Methodology of NLP in sentiment analysis

The process involves data collection, text preprocessing, feature extraction, sentiment classification using machine learning models, and analysis and interpretation of results. This provides valuable insights for policymakers and organizations to develop effective communication strategies and address public knowledge gaps.

Decision Support Systems: Architecture and Integration with GIS

Overview of DSS architecture

Decision Support Systems (DSS) in the realm of Artificial Intelligence and environmental management have a modern and detailed architecture meant to aid complex decision-making processes. This architecture incorporates a number of components, each of which is critical to the system's effectiveness.

Core components of AI-based DSS

Data management layer

This foundational layer is responsible for the collection, storage, and management of a wide array of data types. For pollution management, it encompasses environmental data from various sources such as satellite imagery, sensor data from monitoring stations, and unstructured data from social media and news sources.

Processing and analysis layer

Here, AI algorithms process and analyze the data. Machine learning models, including deep learning networks, are employed to identify patterns, trends, and anomalies crucial for understanding pollution levels and environmental changes.

Predictive analytics engine

Utilizing the power of AI, this engine forecasts future trends and potential pollution hotspots, enabling a proactive approach to environmental management.

Decision-making and recommendation layer

This layer synthesizes insights from analysis and predictive analytics to provide actionable recommendations, supporting decision-makers in formulating effective strategies.

User Interface (UI)

The UI is the front-end through which users interact with the DSS. It presents data and insights in an accessible and understandable format, often using visualizations.

Real-time data processing

Real-time data processing is a critical feature of AI-based DSS in pollution management. It enables immediate analysis of incoming data, crucial for monitoring sudden pollution events and implementing timely mitigation strategies.

AI-driven recommendations

These recommendations, based on comprehensive data analysis, provide decision-makers with evidence-based suggestions for both short-term responses and long-term strategic planning.

Integration with Geographic Information Systems (GIS)

The integration of DSS with Geographic Information Systems (GIS) enhances their capability in pollution management. GIS offers spatial analysis tools and geographically referenced data, providing a powerful platform for visualizing, analyzing, and managing environmental data.

Benefits of GIS integration

- *Spatial Analysis*: GIS allows for the spatial analysis of pollution data, enabling the identification of geographic patterns and trends.
- *Data Visualization*: Advanced visualization tools in GIS aid in understanding complex data through maps and spatial representations.
- *Enhanced Decision-Making*: The combination of GIS with DSS leads to more informed decision-making by providing a comprehensive view of spatial and temporal aspects of environmental data.

Enhancing Environmental Management through GIS-DSS Synergy

The role of GIS in pollution management

GIS technology captures, stores, analyzes, manages, and presents spatial or geographic data. In pollution management, GIS functions in several key ways:
- *Spatial Data Visualization*: GIS excels at visualizing spatial data, aiding in mapping pollution levels and understanding geographic spread.

- *Geospatial Analysis*: GIS performs spatial analysis such as calculating pollution spread, modelling pollutant dispersion patterns, and assessing impacts on different zones.[1]
- *Data Integration*: GIS integrates various data types, crucial for comprehensive pollution management strategies.

Enhancing DSS with GIS

The integration of GIS with DSS in environmental management enhances both functionality and accuracy. It improves decision-making, enables targeted mitigation strategies, and facilitates real-time monitoring and response, particularly in urban air quality management, water pollution control, and more.

Interdisciplinary Collaboration and Real-World Applications

Collaborative efforts in AI-powered DSS development

The development and implementation of AI-powered Decision Support Systems (DSS) for pollution management showcase the essence of interdisciplinary collaboration. This collaboration brings together AI and machine learning experts, environmental scientists, policymakers, data analysts, GIS specialists, community stakeholders, and NGOs, each contributing their expertise to create effective environmental solutions.

Real-world applications of AI-powered DSS

AI-powered DSS find applications in diverse areas:

- *Urban Air Quality Management*: Monitoring air quality, predicting pollution levels, and informing policy for emission control (Osorio et al. 2011).
- *Water Quality Monitoring*: Tracking pollution sources in water bodies and guiding cleanup efforts (Behmel et al. 2021).
- *Waste Management Optimization*: Aiding in efficient waste collection and recycling initiatives (Fang et al. 2023).
- *Agricultural Pollution Control*: Assisting in sustainable farming practices and minimising agricultural pollution (Mir et al. 2015).

---

[1] The GIS integration involved the use of ArcGIS software for spatial analysis, which helped in mapping pollution hotspots and analyzing their proximity to residential areas.

Case Studies: Demonstrating the Effectiveness of AI-Powered DSS

The practical application of AI-powered Decision Support Systems (DSS) in various environmental contexts is best illustrated through real-world case studies. These examples showcase the adaptability, effectiveness, and practicality of AI-powered DSS in addressing diverse environmental challenges.

Case Study 1: Urban air quality monitoring

Location and context

A major metropolitan city grappling with rising air pollution levels due to increased vehicular traffic and industrial activities.

AI-powered DSS implementation

Implementation of an AI-powered DSS that integrated real-time data from air quality monitoring stations with satellite imagery and traffic data.[2]

Outcomes

The system successfully predicted pollution hotspots and informed policy interventions like traffic regulation in high-pollution zones and stricter emission standards, resulting in improved air quality.

Case Study 2: Water quality management in a river basin

Location and context

A river basin experiencing frequent contamination events, impacting aquatic life and dependent human populations.

AI-powered DSS implementation

Development of an AI-powered DSS to monitor water quality using sensor data and satellite imagery.

Outcomes

Early warnings of contamination events were provided, leading to quicker response and mitigation, and the overall health of the river improved due to targeted cleanup operations and stricter regulations.

---

[2] The AI model used in the urban air quality study was a modified version of the convolutional neural network (CNN), which was specifically trained to identify emission patterns from industrial and vehicular sources.

Case Study 3: Optimizing waste management in a growing city

Location and context

An urban area facing challenges in managing increasing waste output efficiently.

AI-powered DSS implementation

Deployment of an AI-powered DSS to optimize waste collection routes and predict waste generation patterns.

Outcomes

Reduction in fuel consumption and operational costs, better resource allocation, higher recycling rates, and reduced landfill usage were achieved.

Case Study 4: Agricultural pollution control

Location and context

A rural region with extensive agricultural activities leading to pollution due to overuse of pesticides and fertilizers.

AI-powered DSS implementation

Use of an AI-powered DSS to analyze soil and crop data, weather patterns, and farming practices for sustainable agriculture recommendations.

Outcomes

Personalized guidance to farmers led to reduced chemical inputs, sustainable farming practices, and decreased agricultural pollution.

Ethical Considerations and Scalability in AI Application

Ethical Aspects in AI for Environmental Management

The use of AI in environmental management raises important ethical considerations, including data privacy, algorithmic transparency, and impacts on communities. Addressing these aspects is crucial for the responsible and equitable use of AI.

Data privacy

- *Concerns*: Risk of misuse or mishandling of sensitive data collected for AI analysis.
- *Mitigation Strategies*: Implementation of strict data governance policies, data anonymization, and encryption techniques.

Algorithmic transparency

- *Concerns*: The "black box" nature of AI algorithms can lead to trust issues and accountability challenges.
- *Mitigation Strategies*: Development of explainable AI models and regular audits to assess decision-making processes.

Table 1 outlines the differences in efficiency, accuracy, and scope between traditional methods and AI-based approaches in environmental monitoring.

Table 1. Comparison table for AI based & traditional methods in pollution mapping.

Criteria	Traditional Methods	AI-based Methods
Efficiency (time and resources)	More time-consuming and resource-intensive	Efficient due to automation and advanced data processing
Accuracy (precision of measurements and predictions)	Variable accuracy, dependent on manual processes and equipment	Higher accuracy with advanced algorithms and continuous learning
Scope (range and depth of monitoring)	Limited to specific locations or parameters; less dynamic	Broader scope, analyzes large datasets and multiple parameters dynamically

Impact on communities

- *Concerns*: AI-driven decisions can significantly impact communities, potentially affecting marginalized or vulnerable groups.
- *Mitigation Strategies*: Involving community stakeholders in AI system development and conducting impact assessments.

Scalability of AI solutions

Scalability involves adapting and expanding AI technologies to meet diverse environmental challenges across different contexts.

Factors contributing to scalability

- *Adaptability*: The ability of artificial intelligence models to adjust to changing environmental conditions.

- *Cost-Effectiveness*: Providing artificial intelligence solutions at a low cost, particularly in resource-constrained contexts.
- *Interoperability*: The ability of artificial intelligence-based systems to incorporate with modern innovations and data networks is referred to as interoperability.
- *Capacity Building*: Investing in training and capability development, particularly in developing countries.

Potential and Challenges

Potential

Scalable AI solutions offer immense potential for global sustainability efforts, with the ability to be tailored to specific challenges.

Challenges

Ensuring equitable access, avoiding one-size-fits-all solutions, and addressing the digital divide.

Conclusion

AI offers transformative potential for environmental management, but it is imperative to navigate ethical considerations and focus on scalable solutions that are adaptable, equitable, and accessible globally.

Acknowledgments

"I would like to extend my heartfelt thanks to Professor Surya Durbha at IIT Bombay and Dr Arun, IIIT Sri City for their invaluable guidance and mentorship throughout this research. My gratitude also goes to my colleagues in the Environmental AI Research Group for their insightful feedback and collaborative spirit. I am grateful to the Indian Council of Environmental Research for funding this study. Lastly, I would like to thank my family for their unwavering support and encouragement."

References

Arashpour, M. 2023. AI explainability framework for environmental management research. Journal of Environmental Management 342: 118149.

Behmel, S., Damour, M., Ludwig, R. and Rodriguez, M. J. 2021. Intelligent decision-support system to plan, manage and optimize water quality monitoring programs: Design of a conceptual framework. Journal of Environmental Planning and Management 64(4): 703–733.

Chaurasia, K., Kanse, S., Yewale, A., Singh, V. K., Sharma, B. and Dattu, B. R. 2019, December. Predicting damage to buildings caused by earthquakes using machine learning techniques. pp. 81–86. In 2019 IEEE 9th International Conference on Advanced Computing (IACC).

Chaurasia, K., Tarun, U., Sarala, G. V. and Soni, K. 2020a, November. AI based prediction of daily rainfall from satellite observation for disaster management. pp. 176–188. In SPIE Future Sensing Technologies (Vol. 11525).

Chaurasia, K., Neeraj, B., Burle, D. and Mishra, V. K. 2020b, September. Topographical feature extraction using machine learning techniques from sentinel-2A imagery. pp. 1659–1662. *In*: IGARSS 2020–2020 IEEE International Geoscience and Remote Sensing Symposium.

Fang, Bingbing, Jiacheng Yu, Zhonghao Chen, Ahmed I. Osman, Mohamed Farghali, et al. 2023. Artificial intelligence for waste management in smart cities: a review. Environmental Chemistry Letters: 1–31.

Huang, I. B., Keisler, J. and Linkov, I. 2011. Multi-criteria decision analysis in environmental sciences: Ten years of applications and trends. Science of the Total Environment 409(19): 3578–3594.

Jena, A. K., Potru, S. S., Balaji, D. R., Madu, A. and Chaurasia, K. 2021, April. Disaster risk mapping from aerial imagery using deep learning techniques. pp. 319–329. In International Conference on Unmanned Aerial System in Geomatics. Cham: Springer International Publishing.

Kuldeep, Garg, P. K. and Garg, R. D. 2021. Texture-based riverine feature extraction and flood mapping using satellite images. Advances in Remote Sensing for Natural Resource Monitoring, pp. 405–430.

Mir, S. A., Qasim, M., Arfat, Y., Mubarak, T., Bhat, Z. A., Bhat, J. A. et al. 2015. Decision support systems in a global agricultural perspective—a comprehensive review. Int. J. Agric. Sci. 7(1): 403–415.

Osorio, M. A., Torrijos, T., Sánchez, A. and Arroyo, O. 2011, July. Preliminary analysis for an air quality management DSS in the metropolitan valley of Puebla, Mexico. pp. 210–215. *In*: 11th WSEAS International Conference on Wavelet Analysis and Multirate Systems: Recent Researches in Computational Techniques, Nonlinear Systems and Control.

Rayhan, A. 2023. AI and the environment: toward sustainable development and conservation.

Silvestro, D., Goria, S., Sterner, T. and Antonelli, A. 2022. Improving biodiversity protection through artificial intelligence. Nat. Sustain 5: 415–424.

Smith, K. 2013. Environmental Hazards: Assessing Risk and Reducing Disaster. Routledge.

Vinuesa, R., Azizpour, H., Leite, I., Balaam, M., Dignum, V., Domisch, S. et al. 2020. The role of artificial intelligence in achieving the Sustainable Development Goals. Nat. Commun. 11: 233.

Wang, J., Zhao, Y., Balamurugan, P. and Selvaraj, P. 2022. Managerial decision support system using an integrated model of AI and big data analytics. Annals of Operations Research 1–18.

Index

About the Editors

Dr. Kuldeep Chaurasia is an Associate Professor at School of Computer Science Engineering and Technology at Bennett University, Greater Noida, India. He earned his Ph.D. in Geomatics from Indian Institute of Technology Roorkee, and made significant contributions as a Research Scientist at the National Remote Sensing Centre, ISRO, Govt. of India, while working on various national projects. His expertise spans machine learning, deep learning in the geospatial domain, flood mapping, spatial data management, computer networks, and geo-blockchain. Dr. Chaurasia has published over 40 research papers in prestigious international journals and conferences, achieving significant breakthroughs and earning widespread respect in the academic community.

Prof. Pradeep K. Garg serves as a Professor in the Geomatics Engineering Group, Civil Engineering Department, IIT Roorkee. He earned his B.Tech and M.Tech from the University of Roorkee, and Ph.D. from the University of Bristol, UK. He has held key positions, including Head of Civil Engineering at IIT Roorkee and Vice Chancellor of Uttarakhand Technical University. Prof. Garg has published over 300 technical papers, authored several textbooks, and supervised numerous theses, significantly contributing to geospatial research and education.

Dr. Arun P. V. is an Assistant Professor in CSE and Institute Research Coordinator at the Indian Institute of Information Technology, Sri City, Chittoor. His research interests include deep learning, computer vision, explainable image and signal processing, remote sensing, and precision agriculture. He completed his Ph.D. at IIT Bombay, and postdoc at Ben-Gurion University, Israel. He served as a Research Scientist/Assoc Vice-president at Huawei R&D. He has published around 20 high-impact SCI-indexed journals, with about 300 citations. He leads projects funded by DST, ISRO, and other organizations, and consults for industries. He has reviewed top journals, and collaborated with various national and international agencies.

Dr. Yingwei Yan is a GIScientist working on volunteered geographic information, social sensing, and GIS educational research. He is the Director of Taught Graduate/Continuing Education and Training programmes in Applied GIS, and a member of the GIS Unit of the Department of Geography, National University of Singapore (NUS). He obtained his Ph.D. from the Department of Geography, NUS. He is also the Co-chair of the ISPRS Technical Commission Working Group on Human Behaviour and Spatial Interactions.